D0580054

# Extra-Terrestrial Matter

# Extra-Terrestrial Matter

Charles A. Randall, Jr.
Editor

Northern
Illinois
University
Press

Proceedings of a Conference
at Argonne National Laboratory
March 7-8, 1968

Cover — Yerkes Observatory photo. Whirlpool nebula (M51) (Canes Venatici).
Taken by Van Biesbroeck with the 82-inch reflector, MacDonald Observatory.

Published by the Northern Illinois University Press, DeKalb, Illinois
Manufactured in the United States of America

# Foreword

The current conference topic is extra-terrestrial matter. A conference of this sort and magnitude doesn't just happen. A lot of people do a lot of hard work. I would like to acknowledge the committee who assisted me in doing the necessary spade work and making all of the arrangements: D. L. Bushnell, M. S. Lougheed, C. A. Randall. We had very valuable assistance from several of the Argonne scientists, in particular, L. H. Fuchs, J. Kastner, and G. W. Reed. We acknowledge the assistance with innumerable details that must accompany such a venture as this, generously given by Lou Dini, M. D. Peterson, Miss Dorothy Carlson and others whose names I don't know but I'm sure there were many behind-the-scenes operations going on in our behalf. It's my pleasure at this time to introduce Robert Duffield, recently appointed Director of Argonne National Laboratory, who will extend greetings for the laboratory.

*A. F. Foster*
*University of Toledo*

# Welcoming Remarks

The conference is sponsored jointly by CSUI, by AMU, by Argonne, and by the AEC. We feel that this program of cooperation between the educational institutions and Argonne Laboratory is a very useful and beneficial thing for us and, we hope, also for the educational institutions that are involved. This kind of cooperation has been going on for some time, of course; however, since I haven't been here for several years but only for four months, I can't claim any personal knowledge of the early history of this cooperation. I do want to express my personal enthusiasm for it and say that as far as the laboratory is concerned, we think it is a fine activity. The opportunity that the conference provides to have informal conversations, discussions and research planning among the members of the educational institutions and the Argonne Laboratory is particularly important. We hope that there will be more of it rather than less.

Fermi is reputed to have said that atomic energy was great stuff and that it would be particularly nice if it could do things like curing the common cold. We have made little progress on that, but we are optimistic about many other things that nuclear energy can do. It is interesting to me to see that it is doing things such as telling us the structure of the lunar surface. Maybe it will even help us to get to the moon someday and find out directly what it looks like. We are happy that, though most of our activities here at the laboratory are concerned with terrestrial problems, we do have the ability to accomplish some things which are related to extra-terrestrial problems. I hope that when you have another conference on this subject in a year or two we can point to even more results that are relevant to this new and exciting field.

We are glad to provide the physical facilities for your comfort and make what arrangements we can to help you have this conference. I hope that you will find it is a stimulating environment for your meetings and discussions.

*Robert Duffield*
*Director*
*Argonne National Laboratory*

# Editor's Preface

The CSUI-ANL Conference on Extra-terrestrial Matter is the third technical conference arranged and sponsored by Central States Universities, Inc., in cooperation with Argonne National Laboratory, the Division of Nuclear Education and Training of the United States Atomic Energy Commission, and the Associated Midwest Universities. With the advent of successful landings on the moon, it was felt that it was an appropriate time to review discoveries in geophysics, geochemistry and astrophysics at distances greater than one earth radius.

To give authority to the review papers, investigators who are active in the areas of interest to the Central States universities were invited to participate.

The program was organized to provide ample time for presentations followed by informal open discussion. Fourteen papers on research in extra-terrestrial matter and related topics were presented. Over one hundred scientists attended and participated in the discussions.

The book is divided into six sections according to the interests of the member universities of CSUI:

Section One: *Meteorites and Tektites*. Recent investigations into the composition of stony meteorites, which normally includes the chondrites, achondrites and stony-irons, are reviewed. One recent challenging problem is the presence of organic matter in meteorites. Consequently, it was fitting to include new evidence on the origin of such material. This session concluded with a review of the most recent data upon which we can base hypotheses for the origin of tektites.

Section Two: *Lunar Science*. Out of many possible exciting problems arising from the current studies of the moon, three were chosen as excellent examples of methodology and recent results. Work on the lunar regolith was reviewed as an example of new astrogeology. Chemical analysis of the lunar surface by methods developed at the University of Chicago and Argonne National Laboratory were presented. The study of the effects of solar protons incident upon the lunar surface through the use of luminescence and chemical effects were summarized as an appropriate conclusion to this session.

Section Three: *Cosmic Rays.* This section was presented during an evening session and comprises a comprehensive survey of the chemical composition of galactic and solar cosmic rays.

Section Four: *Cosmic Rays and Atmospheric Neutrons.* The emphasis in this section is upon interactions closer to the earth. The most recent work on the electron component of cosmic rays was presented, followed by a summary of cosmic ray neutron studies. The problem of neutron exposure encountered by personnel in the new supersonic transport was discussed. The complex interactions of cosmic ray neutrons at the air-soil and air-water interfaces are presented. The session concluded with a summary of carbon-14 production by cosmic rays in the atmosphere.

Section Five: *Comets and Interstellar Dust.* A review of our present knowledge of interstellar dust is presented. Interpretations of the material in cometary nuclei are discussed.

Section Six: *Summary.* The conference is summarized by a transcript of the panel discussion made up of John A. O'Keefe, Goddard Space Flight Center; Eugene Shoemaker, U.S. Geological Survey; James H. Patterson, ANL-CHM; and Charles A. Randall, Ohio University.

Although some manuscripts were supplied by the authors, the major portion of this book is taken from taped transcripts of the conference. The editor is solely responsible for any errors which occur in such a procedure. The atmosphere of conversational informality that occasionally appears in the session reports is one of the charming features of a small conference, for the editor is reluctant to preside over major surgery which would totally suppress the personalities of the speakers in transforming the lectures into more formal papers. After all, it is a property of many large scientific meetings that not infrequently more scientific information is informally exchanged in the corridors than in the formal sessions. The question and answer sessions which followed appear at the end of each paper insofar as they can be understood upon listening to the tapes.

The editor expresses his sincere thanks to the many individuals who assisted him during the two days of the conference and throughout the preparation of the manuscript.

*C. A. Randall*
*Athens, Ohio*

# *President's Preface*

It is with a great deal of pride that the Central States Universities, Inc., formally presents the proceedings of its recent conference on Extra-terrestrial Matter. With this publication, the fifteen-university organization founded in 1964 initiates another important activity in its over-all cooperative program involving the various scientific disciplines.

The immediacy of the topic chosen for this conference reflects the interest of the various science departments of the member CSUI institutions in current developments in a field which will be of increasing importance and concern as man seeks to gain a foothold in space. As in the two previous technical conferences sponsored by CSUI, an attempt has been made, by careful selection of participants and topics, to provide information of value to as broad spectrum of academic disciplines and interests as may be feasible. Accordingly, it is anticipated that these proceedings will have wide appeal, not only to scientists who are actively engaged in work directly related to the topics selected but also to other scientists and laymen who have either an active interest in the progress of man's conquest of space or are merely curious about the topic. It is hoped that this volume, as such, will provide a significant addition to the literature of the subject.

CSUI would like to take this opportunity to express its appreciation to those individuals and groups not directly affiliated with its member universities whose efforts, encouragement, and help have played such a large part in the eventual conference. Particular thanks are extended to the speakers who have so kindly contributed their energy and talents to providing the subject matter of the conference; to the Argonne National Laboratories which, as a principal co-sponsor of the conference, not only made available its exemplary facilities for the meetings but also provided some of the speakers and was otherwise of inestimable assistance in planning and implementing both the scientific and the non-scientific aspects thereof; to the Division of Nuclear Education and Training of the United States Atomic Energy Commission for providing funds and encouragement for the conference, and to the Associated Midwest Universities, whose co-sponsorship of the conference indicated its widespread value to the entire educational communities of the Midwest, particularly those with graduate programs.

Finally, thanks are especially due to the members of the joint CSUI-ANL conference committee which planned, organized, and implemented the conference itself.

It is a reasonable expectation that this volume is the forerunner of similar proceedings of future CSUI technical conferences, and, as such, that it marks a special milestone in the progress of CSUI activities.

*Hugh F. Henry, President*
*Board of Directors*

*Central States Universities, Inc.*

# Contents

# *Tables*

# *Figures*

# I. Meteorites and Tektites

# Composition of Stony Meteorites

BRIAN MASON

*Abstract: Studies of meteorites suggest that they are samples of material which crystallized at a very early stage in the development of the solar system. It is proposed that in the pre-planetary stage of the solar system they crystallized shortly after the last stage in nuclear synthesis. The mineralogy and structure of the chondrites indicate a variety of formation conditions. The structure of enstatite chondrites indicates crystallization at temperatures as high as 1500° C., whereas the hydrates in carbonaceous chondrites show formation under cold and humid conditions. The general conclusion is that these meteorites represent samples from different regions of an ancestral solar nebula where the temperature ranged from very hot near the center to quite cold on the margins. The chondrites are an inhomogeneous aggregation of solid materials which may have never been completely melted, in contrast with the achondrites and stony-irons which appear to have originated in asteroids which grew large enough to melt or at least partially melt. Differentiation could have occurred by gravitational separation and fractional crystallization. Probably the chondrites came from bodies which never grew larger than about 50 km in diameter, whereas the achondrites and stony-irons came from rather large asteroids.*

INTRODUCTION

The study of meteorites is an amateur sport, and we are all really amateurs. I don't know anybody in this field who took a course on meteorites in college. We get into meteorite research from some other field such as astronomy, geology, chemistry, physics, or these days sometimes even biology. I started out as a geologist.

By happy chance I have been curator in charge of two of the world's finest meteorite collections, formerly that of the American Museum of Natural History in New York, and at present the collection in the Smithsonian Institution in Washington. So, to me, meteorites are extra-terrestrial rocks, to be studied much the same way as terrestrial rocks.

Research on extra-terrestrial rocks has its frustrating aspects, however. Whereas the earthbound geologist enjoys an infinite amount of material, I am limited to what has randomly fallen and has been randomly collected over the past century and a half. Meteorite falls are rare and unpredictable events. Perhaps this is just as well; otherwise more of us might suffer the unique experience of Mrs. Hodges of Sylacauga, Alabama, who was resting in her living room on a November day in 1954 when an 8-lb meteorite crashed through the ceiling, bounced off her radio and hit her on what was delicately reported as her upper thigh. Mrs. Hodges thereby became the first person to have been injured by a known extra-terrestrial object.

Incidentally, I spent last week in Denver investigating the first recovered meteorite fallen in the United States since 1961. This illustrates one fundamental problem in meteorite research—the rarity of the raw material. Last week I got a little package in the mail with what is probably the shortest letter on record from a firm called Nationwide Papers: "Dear Mr. Mason: This stone came through our roof last August. You might be interested." I opened it up and there was a beautiful little 200-g meteorite. This meteorite, incidentally, was discovered when the employees found water pouring down through a hole in the roof; under that hole was this little meteorite.

CLASSIFICATION

Table 1 shows the classes of meteorites and the number of observed falls in each class. One can classify meteorites as one classifies terrestrial rocks— by their mineralogy and their structure. The basic broad classifications are probably familiar: irons, stony-irons, and stones. The stones are usually divided into two groups: the chondrites and the achondrites, because they're very, very different. Chondrites are so named because they have small, round objects in them called chondrules, and these small round objects are like little round seeds up to about a millimeter in diameter made up of silicate minerals; chondrules are exclusively meteoritic in origin. Chrondrules have never been seen in any terrestrial rock, neither sedimentary, igneous, nor metamorphic. I suggest that chondrules formed by some rather exotic process; hence, the deep significance of division of stones into chondrites and achondrites. The achondrites don't have chondrules, and their structure is rather like some terrestrial igneous rocks. If we were starting off fresh in the classification of meteorites, I think we'd classify all meteorites into chondrites and non-chondrites.

**Table 1. The Classification of Meteorites. (The numbers in parentheses are the numbers of observed falls in each class.)**

| Group | Class | Principal Minerals |
|-------|-------|--------------------|
| Chondrites | Enstatite (11) | Enstatite, nickel-iron |
| | Olivine-bronzite (227) | Olivine, bronzite, nickel-iron |
| | Olivine-hypersthene (303) | Olivine, hypersthene, nickel-iron |
| | Carbonaceous (31) | Serpentine, olivine |
| Achondrites | Aubrites (8) | Enstatite |
| | Diogenites (8) | Hypersthene |
| | Chassignite (1) | Olivine |
| | Ureilites (3) | Olivine, clinobronzite, nickel-iron |
| | Angrite (1) | Augite |
| | Nakhlite (1) | Diopside, olivine |
| | Eucrites and howardites (40) | Pyroxene, plagioclase |
| Stony-irons | Pallasites (2) | Olivine, nickel-iron |
| | Siderophyre (1) (Find) | Orthopyroxene, nickel-iron |
| | Lodranite (1) | Orthopyroxene, olivine, nickel-iron |
| | Mesosiderites (6) | Pyroxene, plagioclase, nickel-iron |
| Irons | Hexahedrites (7) | Kamacite |
| | Octahedrites (32) | Kamacite, taenite |
| | Ni-rich ataxites (1) | Taenite |

Achondrites, stony-irons, and irons can be understood fundamentally as derived by various processes that are familiar in terrestrial rocks—by melting, by fractional crystallization, and by differentiation of liquid and solid. One can derive these groups of meteorites from the chondrites.

The chondrites seem to be more primitive than any of the other types of meteorites; also, the numbers of observed falls show that the chondrites, especially two groups, the olivine bronzite and the olivine hypersthene chondrites, are far more abundant than any other meteorites. Mrs. Hodges was hit by a bronzite chondrite. The one in Denver was a hypersthene chondrite. Widespread informal opinion gives the impression that the irons are very abundant. Irons are abundant because they last a long time after they've fallen on the ground, and they're easily recognized. In actual number of falls the irons are rather rare.

One interesting feature of the different classes of meteorites is that some of them are very poorly inhabited; the classification shows only one angrite, only one nakhlite, one siderophyre, and one lodranite, which suggests to me that, while we've been collecting meteorites for the last century and a half, perhaps we haven't seen all the different types of meteorites yet. I'm always hopeful that someday I'll get a type of meteorite which has never been seen before. It is quite possible.

While the over-all mineral compositions can be understood in general terms, there are curious hiatuses in meteoritic composition; thus almost certainly there are meteorites with mineralogy that has not yet been seen. One interesting group of which you've undoubtedly heard (if you read such scientific journals as *Time* and *Life*) is the carbonaceous chondrites; these contain organic compounds and some people think they contain extra-terrestrial microfossils. It is very curious that they're made up largely of the mineral serpentine. Serpentine is a mineral which contains 14 per cent combined water, which indicates that some part of outer space is cold and humid. Serpentine is unstable above $500°$ C.

The achondrites are essentially metal and sulfide free; the irons are simple and straightforward. The stony-irons divide into pallasites and mesosiderites. The pallasites contain rather a lot of $MgO$ and less $FeO$, whereas the mesosiderites contain rather more $FeO$ and less $MgO$. Achondrites are usually divided chemically into the calcium-poor and the calcium-rich achondrites, although there is actually a fairly continuous sequence. Here again it is found that as the amount of $CaO$ increases so does the amount of $FeO$. For example, the aubrites are essentially pure enstatite with no metal and no calcium silicates.

The chondrites can be divided into four classes, which show a stepwise sequence of chemical and mineralogical composition. The carbonaceous chondrites contain practically no metal and most of the iron is combined in silicates. The hypersthene and bronzite chondrites contain iron combined in silicates and also present as metal and sulfide. In the enstatite chondrites the silicates are essentially iron-free, all the iron being present as metal and sulfide.

METEORITE MINERALOGY

The first comprehensive account of meteorites was by Tschermak in 1885. He described sixteen minerals known to be found in meteorites. The miner-

**Table 2. Twenty-six Minerals Reported in Meteorites through 1955.**

| Mineral | Formula | Tschermak 1885 | Farrington 1915 | Heide 1934 | Krinov 1955 |
|---|---|---|---|---|---|
| Kamacite | (Fe, Ni) | x | x | x | x |
| Taenite | (Fe, Ni) | x | x | x | x |
| Copper | Cu | | | | x |
| Diamond | C | | x | x | x |
| Graphite | C | x | x | x | x |
| Moissanite | SiC | | x | x | x |
| Cohenite | $Fe_3C$ | | x | x | x |
| *Schreibersite | $(Fe, Ni)_3P$ | x | x | x | x |
| *Osbornite | TiN | x | x | x | x |
| Troilite | FeS | x | x | x | x |
| *Oldhamite | CaS | x | x | x | x |
| *Daubreelite | $FeCr_2S_4$ | x | x | x | x |
| *Lawrencite | $(Fe, Ni)Cl_2$ | | x | x | x |
| Magnesite | $(Mg, Fe)CO_3$ | x | x | | x |
| Calcite | $CaCO_3$ | | | | x |
| Magnetite | $Fe_3O_4$ | x | x | x | x |
| Chromite | $FeCr_2O_4$ | x | x | x | x |
| Quartz | $SiO_2$ | | x | x | x |
| Tridymite | $SiO_2$ | x | x | x | x |
| Chlorapatite | $Ca_5(PO_4)_3Cl$ | | x | x | x |
| Merrillite (Whitlockite) | $Ca_3(PO_4)_2$ | | | x | x |
| Olivine | $(Mg, Fe)_2SiO_4$ | x | x | x | x |
| Orthopyroxene | $(Mg, Fe)SiO_3$ | x | x | x | x |
| Clinopyroxene | $(Ca, Mg, Fe)SiO_3$ | x | x | x | x |
| Plagioclase | $(Na, Ca) (Al, Si)_4O_8$ | x | x | x | x |
| Serpentine | $(Mg, Fe)_6Si_4O_{10}(O, OH)_8$ | | | | x |

*Not known to occur in terrestrial rocks.

alogy of meteorites grew very slowly. Krinov in 1955 presented a table of twenty-six, of which five are unknown in terrestrial rocks. (A summary of the development of meteorite mineralogy up to 1955 is given in Table 2). The common and abundant minerals of meteorites are kamacite and taenite, the alpha and gamma forms of nickel-iron; troilite, iron sulfide; the magnesium-iron silicates olivine and pyroxene, and serpentine in carbonaceous chondrites; and plagioclase, sodium-calcium aluminosilicate. Some accessory minerals are highly significant in elucidating conditions of crystallization of meteorites. For example, tridymite has been found in small amounts in several classes of meteorites. Tridymite is a form of $SiO_2$ which is not stable above 3 kilobars pressure; hence the presence of tridymite in a meteorite indicates crystallization at low pressure, i.e., in a small parent body—evidence against

**Table 3. Thirty-three Additional Minerals Identified in Meteorites Since 1955.**

| Mineral | Formula | Reference |
|---------|---------|-----------|
| Sulfur | S | DuFresne and Anders, 1962 |
| *Perryite | $Ni_3Si$ | Fredriksson and Henderson, 1965 |
| Mackinawite | FeS | El Goresy, 1965 |
| Pyrite | $FeS_2$ | Ramdohr, 1963 |
| Sphalerite | ZnS | Ramdohr, 1963 |
| Alabandite | (Mn, Fe)S | Dawson et al., 1960 |
| *Niningerite | (Mg, Fe)S | Keil and Snetsinger, 1967 |
| Pentlandite | $(Fe, Ni)_9S_8$ | Sztrokay, 1960 |
| Cubanite (Chalcopyrrhotite) | $CuFe_2S_3$ | Ramdohr, 1963 |
| Chalcopyrite | $CuFeS_2$ | Ramdohr, 1963 |
| Valleriite | $CuFeS_2$ | Ramdohr, 1963 |
| *Djerfisherite | $K_3CuFe_{12}S_{14}$ | Fuchs, 1966 |
| *Gentnerite | $Cu_8Fe_3Cr_{11}S_{18}$ | El Goresy and Ottemann, 1966 |
| Cristobalite | $SiO_2$ | Dawson et al., 1960 |
| Rutile | $TiO_2$ | Buseck and Keil, 1966 |
| Ilmenite | $FeTiO_3$ | Yudin, 1956 |
| Spinel | $MgAl_2O_4$ | Sztrokay, 1960 |
| *Sinoite | $Si_2N_2O$ | Andersen et al., 1964 |
| Dolomite | $CaMg(CO_3)_2$ | DuFresne and Anders, 1962 |
| Gypsum | $CaSO_4 \cdot 2H_2O$ | DuFresne and Anders, 1962 |
| Epsomite | $MgSO_4 \cdot 7H_2O$ | DuFresne and Anders, 1961 |
| Bloedite | $Na_2Mg(SO_4)_2 \cdot 4H_2O$ | DuFresne and Anders, 1962 |
| *Farringtonite | $Mg_3(PO_4)_2$ | DuFresne and Roy, 1961 |
| Graftonite | $(Fe, Mn)_3(PO_4)_2$ | Olsen and Fredriksson, 1966 |
| Sarcopside | $(Fe, Mn)_3(PO_4)_2$ | Olsen and Fredriksson, 1966 |
| *Panethite | $Na_2Mg_2(PO_4)_2$ | Fuchs, et al., 1966 |
| *Brianite | $Na_2MgCa(PO_4)_2$ | Fuchs, et al., 1966 |
| *Stanfieldite | $Mg_3Ca_4Fe_2(PO_4)_6$ | Fuchs, 1967 |
| Zircon | $ZrSiO_4$ | Marvin and Klein, 1964 |
| Ureyite | $NaCrSi_2O_6$ | Frondel and Klein, 1965 |
| *Merrihueite | $(K, Na)_2Fe_5Si_{12}O_{30}$ | Dodd et al., 1965 |
| *Roedderite | $(K, Na)_2Mg_5Si_{12}O_{30}$ | Fuchs et al., 1966 |
| Richterite | $Na_2CaMg_5Si_8O_{22}(OH, F)_2$ | Olsen, 1967 |

* Not known to occur in terrestrial rocks.

the theory that meteorites are fragments of a disrupted planet. A notable feature of the over-all mineralogy of meteorites is the absence of phases indicative of high pressures (i.e., large parent bodies); the origin of diamond in the Canyon Diablo iron has been plausibly ascribed to the shock of impact with the earth, which formed Arizona's Meteor Crater, and the presence of diamond in the small class of ureilites appears to be due to extra-terrestrial shock effects.

Table 3 shows the minerals that have been described since 1955. The mineralogy of meteorites has doubled since 1955. The names Olsen and Fuchs appear here rather frequently; in fact, it's very difficult to keep up with the

**Table 4. Analyses of Meteorites Belonging to Different Chondrite Groups.**

|  | 1 | 2 | 3 | 4 | 5 |
|---|---|---|---|---|---|
| Fe | 23.70 | 15.15 | 6.27 | 4.02 | 0.00 |
| Ni | 1.78 | 1.88 | 1.34 | 1.43 | 0.00 |
| Co | 0.12 | 0.13 | 0.046 | 0.09 | 0.00 |
| FeS | 8.09 | 6.11 | 5.89 | 5.12 | 3.66(S) |
| $SiO_2$ | 38.47 | 36.55 | 39.93 | 34.82 | 27.81 |
| $TiO_2$ | 0.12 | 0.14 | 0.14 | 0.15 | 0.08 |
| $Al_2O_3$ | 1.78 | 1.91 | 1.86 | 2.18 | 2.15 |
| MnO | 0.02 | 0.32 | 0.33 | 0.20 | 0.21 |
| FeO | 0.23 | 10.21 | 15.44 | 24.34 | 27.34(b) |
| MgO | 21.63 | 23.47 | 24.71 | 23.57 | 19.46 |
| CaO | 1.03 | 2.41 | 1.70 | 2.17 | 1.66 |
| $Na_2O$ | 0.64 | 0.78 | 0.74 | 0.69 | 0.63 |
| $K_2O$ | 0.16 | 0.20 | 0.13 | 0.23 | 0.05 |
| $P_2O_5$ | trace | 0.30 | 0.31 | 0.20 | 0.30 |
| $H_2O$ | 0.34 | 0.21 | 0.27 | 0.10 | 12.86 |
| $Cr_2O_3$ | 0.23 | 0.52 | 0.54 | 0.58 | 0.36 |
| NiO | 0.11 | — | — | 0.00 | 1.53 |
| CoO | — | — | — | 0.00 | 0.07 |
| C | 0.32 | — | 0.03 | 0.19 | 2.48 |
|  | 99.89(a) | 100.29 | 99.67 | 100.08 | 101.01(b) |

1. Enstatite chondrite (Daniel's Kuil; Prior, 1916, p. 14); (a) includes CaS 0.86, $Cr_2S_3$ 0.29.
2. Olivine-bronzite chondrite (Oakley; Wiik, 1956, p. 280).
3. Olivine-hypersthene chondrite (Kyushu; Mason and Wiik, 1961, p. 274).
4. Olivine-pigeonite chondrite (Warrenton; Wiik, 1956, p. 280).
5. Carbonaceous chondrite (Mighei, Wiik, 1956, p. 280); (b) Wiik reported all S as FeS, but it is given here as S, and the corresponding Fe is reported as FeO; Melikoff and Krschischanovsky (1898) found 3.69% total S in Mighei, divided as follows: FeS 0.46%; S (free) 3.19%; $SO_3$ (sulphate), 0.85%; $S_2O_2$ (thiosulphate), 0.12%. Wiik also reports 0.36% loss on ignition, mainly organic matter.

Chicago school. This table was made up for a lecture a year ago, and it's already out of date; for example, I've just been informed this morning that they have another one now, krinovite.

CHONDRITES

Let us return to the chondrites, the most abundant meteorites. I am reluctant to say that any meteorite is primitive material but, as Orwell wrote in *Animal Farm* that some animals are "more equal" than others, so evidence is piling up that at least the chondrites are "more primitive" than the other types of meteorites. We still don't have any idea of just how primitive they are, but one very important characteristic of the chondrites is that their over-all chemical composition, both in major elements and in minor and trace elements, shows a good correlation with the abundance of the elements as seen in the solar spectrum. We believe that the chondrites, especially the carbonaceous chondrites, provide us with an adequate average sample of the non-volatile material in the universe. There is a constant cross-correlation between the elemental abundances as predicted by theories of nuclear synthesis and the elemental abundances which we find in chondrites. This has been a very important field of meteorite research over the last ten to fifteen years. Table 4 shows analyses of five different types of chondrites. One interesting feature which emerges is that the presence of free nickel-iron follows a sequence— a lot of it in the enstatite chondrites and bronzite chondrites, not so much in the hypersthene or the pigeonite chondrites, and none at all in the carbonaceous chondrites. Along with that, the FeO content varies inversely as the metal content. In the chondrites, magnesium, silicon, iron, and oxygen make up over 90 per cent of the material both in terms of numbers of atoms and in percentages. Notice the high combined water content in the carbonaceous chondrites. It is there partly as serpentine minerals, partly as organic compounds.

One of the important reasons why geologists have always been interested in meteorites is that meteorites provide a type of rock which we believe exists in the interior of the earth. Such material, or a modified version of it, seems to make up a large part of the earth. There has always been this assumed correlation between the study of meteorites and the study of the internal composition and structure of the earth. The composition can be expressed in many different ways. The composition or the variation in composition of meteorites is illustrated quite well by Table 5. One specific method which is commonly used by chemists and physicists in comparing meteoritic material with terrestrial material and solar material is to normalize to ten thousand or a million atoms of silicon; then the variation in composition can be seen in terms of numbers of atoms. This procedure, of course, shows that oxygen, iron, silicon, and magnesium constitute by far the largest amount of the commonest elements. I learned one of the facts that is most impressive when I first got interested in meteorites. This was that the amounts of magnesium and silicon are almost the same in all the chondrites. This puzzled me for a while before I read a bit more on nuclear physics and elemental abundances, and then I realized that magnesium is element 12 and silicon is 14, so one would expect them to be approximately equal in abundance in primitive material.

In the olivine-hypersthene chondrites the relative amount of iron is somewhat less than six thousand atoms, whereas in the other classes it is about eight to nine thousand. The oxygen-silicon ratio is almost exactly three in

**Table 5. Comparison of the Average Composition of the Different Classes of Chondrites, in Units of Atoms/10,000 Atoms Si.**

|    | 1 | 2 | 3 | 4 | 5 | 6 |
|----|--------|--------|--------|--------|--------|--------|
| O  | 29,670 | 34,130 | 34,670 | 40,830 | 55,110 | 76,860 |
| Fe | 7630   | 8175   | 5815   | 8179   | 8389   | 9036   |
| Si | 10,000 | 10,000 | 10,000 | 10,000 | 10,000 | 10,000 |
| Mg | 7894   | 9651   | 9406   | 10,630 | 10,430 | 10,610 |
| S  | 2213   | 989    | 1041   | 1192   | 2253   | 4991   |
| Ni | 412    | 459    | 282    | 418    | 450    | 474    |
| Ca | 350    | 521    | 510    | 743    | 719    | 681    |
| Al | 610    | 743    | 708    | 943    | 908    | 842    |
| Na | 464    | 457    | 464    | 345    | 411    | 647    |
| Cr | 79     | 92     | 90     | 125    | 120    | 120    |
| Mn | 53     | 78     | 74     | 53     | 63     | 84     |
| P  | 69     | 58     | 58     | 76     | 90     | 122    |
| Co | 26     | 25     | 15     | 21     | 22     | 23     |
| K  | 35     | 34     | 35     | 37     | 28     | 35     |
| Ti | 16     | 21     | 22     | 34     | 22     | 23     |

1. Enstatite chondrites.
2. Olivine-bronzite chondrites (from Mason, 1965).
3. Olivine-hypersthene chondrites (from Mason, 1965).
4. Olivine-pigeonite chondrites (average of analyses by H. B. Wiik, in Mason, 1963).
5. Carbonaceous chondrites, Type II (average of analyses by H. B. Wiik, in Mason, 1963).
6. Carbonaceous chondrites, Type I (average of analyses by H. B. Wiik, in Mason, 1963).

the enstatite chondrites, because they are made up essentially of enstatite, $MgSiO_3$. In the olivine-bronzite, olivine-hypersthene, and olivine-pigeonite chondrites this ratio is between three and four to one, because they consist largely of varying amounts of pyroxene, $(Mg,Fe)SiO_3$, and olivine, $(Mg,Fe)_2 SiO_4$. Oxygen is high in Type I and II carbonaceous chondrites because a significant amount of combined water is present and that pushes up the oxygen (and hydrogen). However, the most remarkable fact about the chondrites is that, although they vary considerably in mineralogy and structure, the over-all elemental composition is rather uniform.

Among the non-carbonaceous chondrites, the mineralogical composition is closely correlated with the chemical composition, especially the ratio of iron to magnesium in the silicate minerals, as illustrated in Figure 1-1. The amount of olivine decreases, and the amount of pyroxene increases in passing from the olivine-pigeonite chondrites (100 MgO:MgO+FeO=60) through the olivene-hypersthene and olivine-bronzite chondrites to the enstatite chondrites (100 MgO : MgO + FeO = 100). Oligoclase, which is sodium-rich plagioclase, remains rather uniform throughout, about 10 per cent. There is always about 5 per cent troilite. Free nickel-iron increases from less than 5 per cent in the olivine-pigeonite chondrites to 15-25 per cent in the enstatite

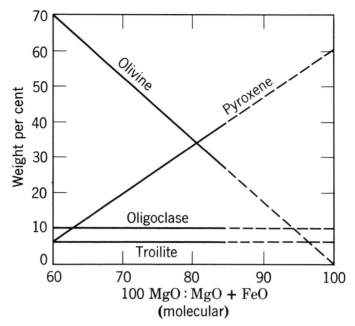

**Fig. 1-1. Schematic representation of the variation of mineralogical composition with chemical composition in the chondritic meteorites; the dashed lines represent the hiatus between the olivine-bronzite and the enstatite chondrites (after Mason, 1960B). Reprinted, by permission, from Mason, *Meteorites*, John Wiley & Sons, 1962.**

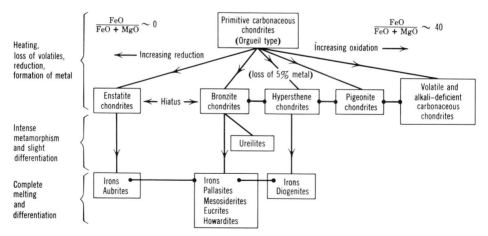

**Fig. 1-2. Genetic relationships between the principal groups of meteorites. (Ringwood, 1961a.) Reprinted, by permission, from Mason, *Meteorites*, John Wiley & Sons, 1962.**

chondrites. This is essentially a reduction series. In moving from left to right along the abscissa the process can be formulated as the reduction of iron in the magnesium-iron silicates and its conversion into free metal. One of the puzzling unsolved problems, one of the things not really understood, is the lack of meteorites in which $100 \, MgO : MgO + FeO =$ is between 85 and 100, i.e., between the olivine-bronzite chondrites and the enstatite chondrites.

G. T. Prior of the British Museum, one of the great authorities on meteorites, contributed some of the best chemical analyses back in the 1910-1920 decade. Prior recognized the over-all compositional similarity between different classes of chondrites, and he put forward the idea that, starting with the material of an enstatite chondrite, one might get the other classes of chondrites by a process of oxidation. Other scientists such as Dr. Ringwood in Australia suggested that the process may have gone the other way: starting from carbonaceous chondrites and obtaining the other classes of chondrites by reduction. This scheme is illustrated in Figure 1-2. This genetic diagram starts with the carbonaceous chondrites of the Orgueil type, and the progeny are produced by heating, reduction, and loss of volatiles.

This sort of relationship between the chondrites can be represented very crudely and very generally by means of the chemical equation:

$$Mg_4Fe_2Si_4O_{10}(OH)_8 \longrightarrow 2(Mg,Fe)_2SiO_4 + 2(Mg,Fe)SiO_3 + 4H_2O$$

$$\text{serpentine} \qquad\qquad \text{olivine} \qquad\qquad \text{pyroxene}$$

$$+ \, 2C$$

$$4MgSiO_3 + 2Fe + 2CO_2$$

$$\text{enstatite} \quad\; \text{iron}$$

from the left side, with the composition of serpentine, one can produce the ordinary chondrites by heating to about 500 degrees. The OH is driven out and the anhydrous silicates are produced. In other words, the mineralogy found in the common hypersthene and bronzite chondrites is reproduced entirely. If a reducing agent is present, such as carbon, the iron is pulled out of the silicates, metallic iron is produced, and the carbon dioxide is driven out. This is a very neat, simple scheme of producing from carbonaceous chondrites, hypersthene chondrites, bronzite chondrites, and the enstatite chondrites. Of course, just by looking at the meteorite one can't determine the direction of the arrows in the chemical equation. Perhaps there shouldn't be arrows there at all; maybe this is simply an equivalence. With our present knowledge I would be inclined to replace the arrows by equal signs, because I now believe that it is not a sequential reaction; it appears to be a series of equivalences. This probably represents the different types of chondrites from different parts of the ancestral solar nebula. This simply may be evidence of how the chondrites were taken from the primordial matter whose composition and mineralogy were established under different conditions of temperature and pressure out of what was essentially originally a material rather homogeneous in its elemental composition.

The evidence for this new position emerges when one looks at these compositions in detail. Curious hiatuses appear, as more sophisticated ways of looking at the chemistry are developed. Although the over-all composition of all chondrites is rather similar in elements, the discontinuities are found to be even more significant. One of the problems that first challenged my interest in meteorites was a paper (again from Chicago—Chicago seems to be one of the great places for meteorite research) in 1953 by Professors Urey and Harmon Craig on the composition of stony meteorites. One of their procedures was to analyze the chondrites as a function of the weight per cent of iron in metal form and FeS (reduced iron) and plot it against the weight per cent of oxidized iron, the iron present in the silicate. The data points form a band running diagonally across the diagram. The authors concluded from this diagram that they could distinguish two groups of chondrites, of which one averaged about 21 per cent total iron and the other averaged about 28 per cent total iron. These were designated by L for low iron and H for high iron. I recall two reactions when I studied their data: first, I thought that there was rather large error scattering, and second, I thought the spread was so wide that the "two-group" proposal was not convincing.

Dr. Wiik and I started analyzing meteorites, especially the ones which seemed to be in the wrong place on the Urey-Craig diagram. To my surprise, instead of proving Urey-Craig were wrong (as I had hoped to do, of course), I proved they were right, but they were even more right than they knew. Instead of only two groups, there were certainly four groups, and they were separated by marked gaps, as shown in Figure 1-3. The enstatite chondrites all fall along the ordinate. We now have samples of enstatite chondrites from 20 per cent up to 35 per cent of total iron, so there are both low iron and high iron, even higher than any of the high irons of Urey and Craig. The second group is the bronzite chondrites, represented by triangles in the figure. The gap which appears here is the same as the gap between the enstatite chondrites and the bronzites in Figure 1-1. The bronzite chondrites form a very nice cluster. The hypersthene chondrites are spread out somewhat more. The Type III carbonaceous chondrites (also known as olivine-pigeonite chondrites) form a separate group, and the Type I and II carbonaceous chondrites, which contain little or no metal and FeS, form a group on the abscissa. The evidence is clear that there are certainly four distinct groups of

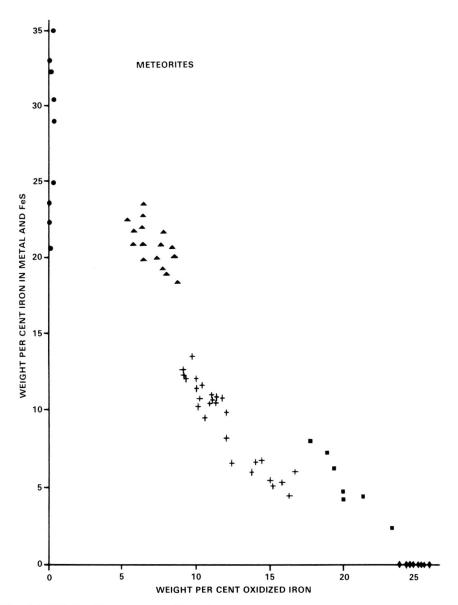

**Fig. 1-3.** Relationship between oxidized iron and iron as metal and sulfide in analyses of chondrites, illustrating the separation into distinct classes and the variation within the classes (● =enstatite chondrites; ▲=bronzite chondrites; +=hypersthene chondrites; ■=carbonaceous chondrites, Type III; ◇ =carbonaceous chondrites, Types I and II).

chondrites separated by hiatuses in their chemistry. All the hypersthene chondrites are members of the L group of Urey and Craig, all other chondrites (except a few enstatite chondrites) belong to their H group.

Other major components show similar grouping (Figure 1-4). Here again,

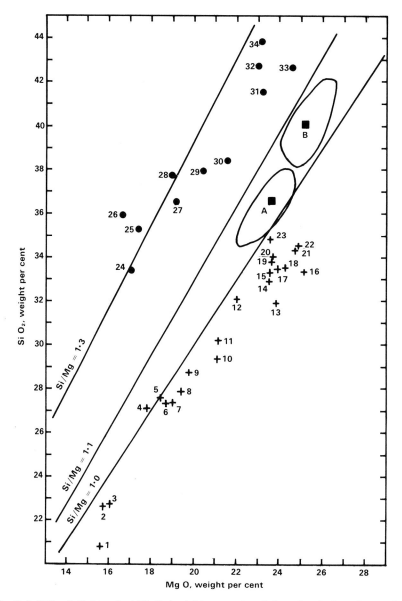

**Fig. 1-4. SiO₂ plotted against MgO (weight percentages) for chemical analyses of chondrites; the diagonal lines are for Si/Mg atomic ratios of 1.0, 1.1, and 1.3. A is the field for 36 analyses of bronzite chondrites, B is the field for 68 analyses of hypersthene chondrites, the black squares being the means for each group. The analyses of carbonaceous chondrites (1-23) and enstatite chondrites (24-34) are plotted individually.**

this remarkable clustering appears. The carbonaceous chondrites all have silicon-magnesium atomic ratios of about 0.95. The area A represents a field of thirty-six of the bronzite chondrites and the area B includes sixty-eight hypersthene chondrites. The points cluster so closely that I simply plotted the averages; their silicon-magnesium ratio has a very uniform value between 1.0 and 1.1. The enstatite chondrites are the solid circles. Their Si/Mg ratio is considerably higher. So at the outset, the chemical compositions of chondrites suggest that they all seem to have come out of the same pot. However, they show marked fractionation.

If the material of the solar system was all uniformly mixed up originally, then the meteorites show some degree of elemental fractionation. One can understand the fractionation of iron. It's not too difficult to fractionate iron, presumably in a dust cloud, if the iron is either as magnetite or in the metal form, because magnetic forces would be quite effective. It's rather more difficult to see how to fractionate silicon and magnesium, because silicon and magnesium are geochemically very similar elements, which almost always form silicates. There is one interesting observation concerning the enstatite chondrites: not all the silicon is there as silicate. Some of the silicon is alloyed with the metal. And the reason for the enhanced silicon value in these meteorites may be that the excess silicon was brought in via metal particles. This is one of the fields which challenges our interest at the present time. How did these fractionations take place? We know that these fractionations occurred at a very early stage in the history of the solar system from such evidence as the fission track dating and from xenon 129. We know chondrules formed within a very short period after nuclear synthesis. So we presume these fractionations took place, also within a short time after nuclear synthesis, on the order of four and a half to five billion years ago.

ACHONDRITES AND STONY-IRONS

Let us now consider the achondrites and stony-irons. Although a wide variety of meteorites is comprised in these, an approach to understanding them can be made in terms of the chemical composition, the principal variables being CaO, FeO, and MgO (Figure 1-5). The enstatite achondrites plot close to the zero-zero mark of this diagram. They are all essentially pure $MgSiO_3$—about the only way to make an enstatite achondrite is to derive it from the enstatite chondrites, by heating and removal of nickel-iron and troilite by liquation and gravitational separation. There is a compositional gap between the enstatite achondrites and the other achondrites and stony-irons, similar to the gap between the enstatite chondrites and the bronzite chondrites. In terms of increasing FeO/FeO + MgO, we have first the pallasites in which this ratio is between 10 and 20, then the hypersthene achondrites, in which

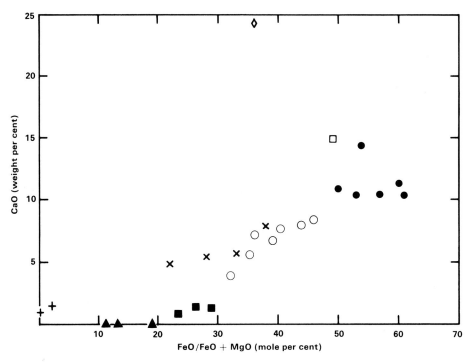

**Fig. 1-5. Plot of CaO (weight per cent) against FeO/FeO + MgO (mole per cent) for the achondrites and stony-irons (+=enstatite achondrites; ▲=pallasites; ■=hypersthene achondrites; ✕=mesosiderites; ○=howardites; ●=eucrites; □=nakhlite; ◇=angrite).**

this ratio is between 20 and 30. In the howardites this ratio is between 30 and 50, and the CaO content increases because of an increase in the amount of calcic plagioclase; mesosiderites are stony-irons in which howardite silicates are mixed with a considerable amount of nickel-iron and troilite. Eucrites have higher FeO/FeO + MgO than howardites, and somewhat higher CaO; the latter is due to somewhat greater amounts of plagioclase, and the presence of pigeonite, a pyroxene richer in calcium than the hypersthene in the howardites. This sequence in chemical and mineralogical composition is essentially identical with that seen in the differentiation of terrestrial ultrabasic and basic igneous rocks, which suggests that similar processes of fractional crystallization of a silicate melt were responsible.

A simple phase diagram helps to explain quite a number of properties of the achondrites. Figure 1-6 is taken from the early work in the Geophysical Laboratory by Andersen in 1915. The system consists of forsterite which is pure magnesium olivine, anorthite (calcium plagioclase), and silica. It is a rather simple phase diagram; iron has been omitted as an extra component. One can substitute iron for some of the magnesium, so instead of dealing with

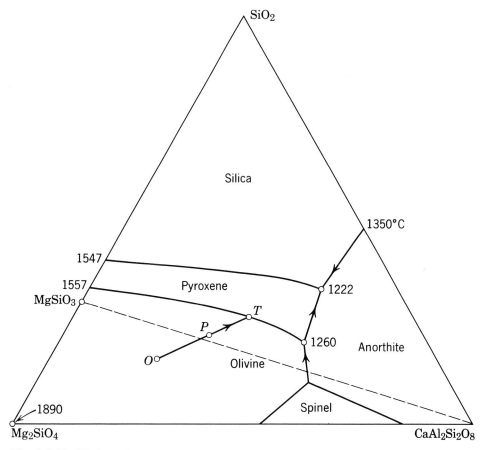

**Fig. 1-6. Equilibrium phase diagram of the system anorthite-forsterite-silica (after Andersen, 1915). For simplicity iron as an explicit component has been omitted. The point 0 is the composition of a silicate in the average chondrite. Reprinted, by permission, from Mason, *Meteorites*, John Wiley & Sons, 1962.**

pure magnesium silicate one is dealing with iron magnesium silicate. After all, magnesium is dominant over iron anyway. The point O is the composition of silicate in the average chondrite, if a melt were made of it. As this melt is cooled and crystallized, olivine would appear first so that the composition of the melt would move from O to T along the line OT. The composition must move to T before there is a phase change to pyroxene. This is because pyroxene is a compound which melts incongruently. So a large amount of olivine can be removed before the melt moves into the pyroxene field. When the melt arrives at this point pyroxene crystallizes out along with the olivine. The olivine along with free nickel-iron from melted chondritic material forms pallasite, and then the system reaches a stage at which pyroxene crystallizes, giving hypersthene achondrites. Once pyroxene begins to crystallize the liquid composition follows the cotectic line to the 1260° point, where plagioclase begins to crystallize, corresponding to the formation of a howardite. The liquid composition then proceeds toward the 1222° point, the lowest temperature on the liquidus, at which point pyroxene, plagioclase, and free silica (tridymite) crystallize together. It is significant that this mineralogical association is characteristic of the eucrites, which thus appear to represent a eutectoid composition.

### ASTEROID MODEL

The whole picture is rather consistent. One can understand the achondrites and the stony-irons. In the details of the over-all picture there are many complications, and one can nit-pick very easily certain specific assumptions, but the over-all picture is fairly consistent. If one can derive the achondrites and the stony-irons by melting up chondrites, there is a certain esthetic appeal. In order to see how this works out, I give you my model asteroid as shown in Figure 1-7. The core of the asteroid is pallasite, nickel-iron and olivine. The pallasite has a density of about 4.3 which corresponds to about 35 per cent of metal. The core mass accounts for 39 per cent of the total mass, and 25 per cent of olivine. This is surrounded by a mantle of hypersthene which accounts for 46 per cent of the mass and 50 per cent of the volume, the density being 3.4. The model is topped off with a crust which will consist of pigeonite and calcium plagioclase. When this is worked out in detail, one gets the hypothetical structure for the asteroid, and the composition corresponds very nicely with the average chondrite, except for one thing: the sodium is off by the factor of the 9. The model must get rid of 90 per cent of the sodium, which is a bit of a puzzle. Maybe that is the origin of the sodium-bearing minerals that Dr. Olsen is finding in iron meteorites.

One interesting feature of this model is that a completely independent piece of evidence from the pallasites is now available. From some very nice work done recently on cooling rates, based on the metal phase of meteorites, it is found that pallasites cooled very slowly indeed, more slowly than any other iron meteorites. This suggests that they were part of a core of an asteroid. This will fit the picture. Even the statistics have a rather interesting consistency. Recall from Table 1 that among the achondrites about forty of them are eucrites and howardites, and about eight of them are hypersthene achondrites; in contrast only two pallasites have been seen to fall. Perhaps all of these meteorites are coming out of one asteroid which has been greatly broken up. There is a great amount of material from the crust, a moderate amount of material from the mantle, and not very much out of the core. The radii of the

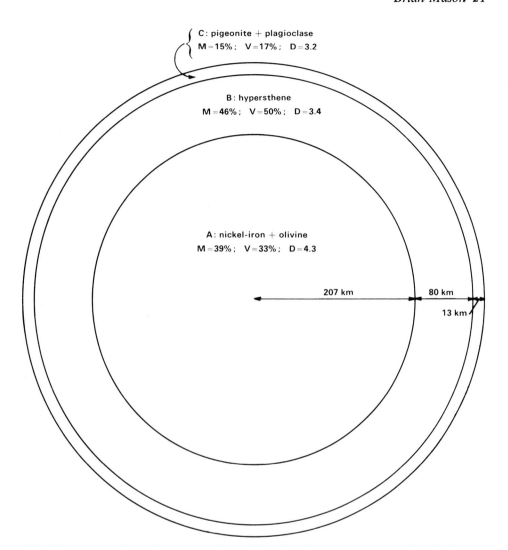

**Fig. 1-7. Cross-section of a hypothetical differentiated asteroid parent of pallasites and achondritic meteorites: nickel-iron, 14 per cent; olivine, 25 per cent; hypersthene, 46 per cent; pigeonite, 9 per cent; plagioclase, 6 per cent; (M=mass, V=volume, D=density).**

|  | SiO$_2$ | MgO | FeO | Al$_2$O$_3$ | CaO | Na$_2$O | Fe | Ni |
|---|---|---|---|---|---|---|---|---|
| **Composition** | 42.7 | 26.2 | 12.7 | 2.3 | 2.0 | 0.1 | 12.5 | 1.5 |
| **Av. chondrite** (omitting FeS) | 41.2 | 26.3 | 13.1 | 2.5 | 2.0 | 0.9 | 12.5 | 1.5 |

layers of the asteroid have been chosen because they fit the cooling rates of the pallasites. From the amount of radioactive elements, from radiation, cooling, etc., it is found that the asteroids cannot have been much more than about 300 or 400 km in diameter for the pallasites ever to have cooled at all. Asteroids cannot be very large; otherwise they wouldn't cool down during all of geological time. For example, Ceres, the largest asteroid known, has a diameter of 380 km.

The chondrites, as far as I can see, are some sort of aggregation of solid material which never completely melted, whereas the achondrites and stony-irons are fragments of asteroids which grew big enough to melt or at least partially melt, so that differentiation by gravitational separation could have occurred. There are, then, these two contrasting groups, the chondrites which have not been through a period of melting and the non-chondrites which have. I propose this is related probably to the size of asteroid. I think the meteorites do come from the asteroids of one sort or another, but the difference between the chondrites and the achondrites is due to a difference in the size of the asteroids. Probably the chondrites come from bodies which never grew very large—never grew more, say, than 50 km in diameter, whereas the achondrites and stony-irons come from rather large asteroids.

## DISCUSSION

(Q) Referring to Table 3, I note the mineral brianite. Other options could have been brian masonite or just masonite.

(A) Mason: When I was approached, I said, "You can't call it masonite because that's already assigned to wallboard."

(Q) As I understand it, one of the questions of interest to the conference is the problem of what research in this area can be done usefully at institutions without tremendous financial resources. I want to mention that Hans Hertz at Goddard has made a study of the orbit of a small asteroid which approaches the asteroid Vesta rather frequently. As a result, he has been able to determine the mass of Vesta and, since there are micrometric measurements on the radius of Vesta, he determined the density of Vesta. Now, you tell me, what about it? What is the density of Vesta?

(A) Mason: Oh, 3.5.

(Q) The mass measurements are pretty good. But the internal evidence is micrometric in rather small bodies. This problem of running up an asteroid orbit is something which should be possible with only a good computer. Many of the universities represented here could do it.

(A) Mason: I quite agree. There is a lot to be done on asteroids. Unfortunately, the astronomers aren't basically interested in meteorites or asteroids. Their view is much farther out into space; I think the solar system has been unfairly dealt with by the astronomers.

(Q) That's right, all you need is a good computer. There are lots of data which can be done several times over for orbits, but isn't it a very simple fact that the study of meteorites requires considerable amounts of material and very complex chemical analyses?

(A) Mason: Well, one of the problems in analyzing meteorites, especially chondrites, where there is a mixture of metal, silicate, and sulphide, is sampling. Take the Denver meteorite, for example. We have 230 gm of it. Now we must analyze this. We'll cut a slice out of it, about 10 gm. We determine its specific gravity, and we'll make a thin section out of it, so we can examine it microscopically. We'll probably grind up the rest; we'll use that for the actual chemical analysis. For about fifteen elements, we'll use about 2 gm. Then the remainder we'll have for supplying to people who want to do trace elements, and so forth, on it. In studying chondrites, it is no great problem to get 10 gm of sample. Occasionally we get meteorites, of course, where we don't even have 10 gm. Then, either we can't do some things or we have to be very selective about what we do. Fortunately, in the case of chondrites, we usually can take 10 gm without depleting the supply too much. It is a problem of curation. Of course, with most meteorites we always have non-homogeneous material. This is a problem because we find metal grains mixed up with silicate grains, and if we're not very careful the metal naturally segregates from the silicate.

(Q) On what basis was the so-called Laplace concept that the asteroids are fragments of a disrupted planet ruled out?

(A) Mason: Well, from the meteorites, everything about their mineralogy indicates they originate in small bodies. The occurrence of tridymite in some meteorites is highly significant, since tridymite has an upper stability limit at about 3 kilobars—this evidence indicates that the parent bodies couldn't have been as big as the moon. One finds 3 kilobars at not too great a depth. The cooling rates are another matter. We now have very good cooling rates on meteorites. The slowest cooling objects we know are pallasites. They cooled about half a degree per million years. There must also be a finite cooling rate, and that's rather difficult to achieve in anything bigger than the largest asteroid. So everything suggests, I think, that, if the meteorites come from the asteroids, the asteroids are not fragments of a disrupted planet. They're simple planetesimals.

(Q) By cooling rates, do you mean while they're in space or after they arrive here?

(A) Mason: While they're in space. The cooling rate is based upon the structure typical of the nickel-iron, the structure where you have the high nickel phase, taenite, intergrown with the low nickel phase kamacite. You can run profiles across these with the microprobe and get the diffusion rates, and from that, you can arrive at the cooling history. These are the rates at which we can presume that this material has gone down through a two-phase region; the originally homogeneous nickel-iron has undergone a solid state transformation into two phases at temperatures of between 500 to 800° C. This must have taken place in the parent body or bodies, whatever they were, before the fragment which we now pick up was ejected into space.

(Q) Would there be complete melting upon entrance into the atmosphere?

(A) Mason: No, the point is that most of the effect which comes from the atmosphere is heating in the form of ablation. The surfaces ablate. It's just as if the meteorites have solved the re-entry problem. The heat only penetrates a few millimeters at the most; otherwise we'd never see a carbonaceous chondrite with its 14 to 20 per cent of water. Even a carbonaceous chondrite has a fusion crust on the outside. The carbonaceous chondrite is like a baked Alaska.

# Origin of Organic Matter in Meteorites  MARTIN STUDIER

*Abstract: A time-of-flight mass spectrometer combined with a gas chromatographic capillary column has been developed. Tests and calibrations with known materials show the gain is three hundred times greater than conventional systems, with a sensitivity of 100 atoms/cc. Results of analyses of organic matter in meteorites suggest that it is not necessary to assume a biogenic origin. New evidence indicates that the Miller-Urey process alone is inadequate to explain the character of the organic matter in meteorites. A remarkable similarity was found between the distribution of hydrocarbons from a synthetic sample and that isolated from a shale believed to be biogenic. The biologically important bases found in DNA and RNA have been produced by heating mixtures of $CO$, $NH_3$, and $H_2$.*

## INTRODUCTION

Dr. Mason spoke of amateurs who work with meteorites. I, too, feel like an amateur. Therefore, I will spend no time reviewing the field since I know so little about it, but I will describe only one field of research in which we have an active interest, in which we have been actively working, and to which I have made some contribution to the solution of the problem. This work has been in cooperation with a group at the University of Chicago headed by

Professor Anders. Dr. Hayatsu has been commuting from Chicago and working with me at Argonne. Also, I'd like to acknowledge Dr. Hayatsu's assistants, Miss Oda and Miss Fuse. At Argonne, I've had the continuous assistance of Mr. Leon Moore; and Mrs. Martin and Mrs. White have helped on a part-time basis.

I shall first describe briefly the apparatus and the techniques we used to investigate this problem and then discuss the application to the problem of organic matter in meteorites. The meteorites are the carbonaceous chondrites, which have a rather high percentage of carbon. The basic tool is the time-of-flight mass spectrometer. Recently we have combined the time-of-flight technique with a very powerful separating tool which has been known for a few years. It is the capillary column, a long capillary tube with a coating of organic matter on the inside, used for separating organic compounds. The column we use is 300 ft long, has a 0.01-in. inside diameter, and is coated with apiazon L, which is a high molecular weight hydrocarbon.

APPARATUS

Table 6 shows the time-of-flight spectrometer as modified for our work. This briefly outlines some of the changes we have made. The original source for this kind of spectrometer was pulsed with only a ¼ per cent duty cycle. Our modifications increased the gain by a factor of 300 by making it a continuous ionization source. Improved resolution resulted. This latter result was a suprising dividend. The sensitivity of the machine operating this way in the ultimate limit is something like $10^{-15}$ torr for nitrogen which corresponds to a steady state of 100 atoms/cc. If there are no interferences, it would be just barely detectable. It has all the sensitivity that can be used. Trace impurities limit sensitivity at much higher levels. When the gain of the machine was increased the multiplier had to see a much larger current, and this would

**Table 6. Improvements in the Time-of-Flight Mass Spectrometer as Modified by Argonne National Laboratory. (The gain was increased by 300 × and the resolution improved.)**

1. CONTINUOUS ION SOURCE
    GAIN INCREASED 300× TO $10^{-15}$ TORR
    IMPROVED RESOLUTION
2. BLANKING GENERATOR
    ELIMINATES SUPERFLUOUS PART OF SPECTRUM
    INCREASES SENSITIVITY
    RETARDS MULTIPLIER FATIGUE
3. AUXILIARY VACUUM SYSTEM
    SAMPLE PREPARATION MONITORED BY TOF
    LOADING OF CAPILLARY COLUMN
4. PAPER CHROMATOGRAPHY — TOF COMBINATION

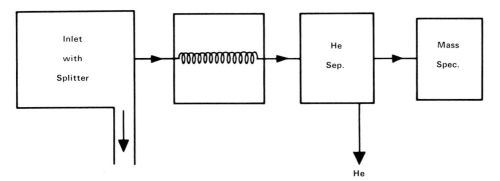

**Fig. 2-1. Sketch of the conventional capillary column gas chromatograph mass spectrometer combination system.**

tend to fatigue and "dirty" the multiplier, so we asked our electronic division to develop what we called a blanking generator, which permits us to blank out any part of a spectrum in which we're not interested. It also increases the sensitivity, because normally the multiplier would have an instantaneous fatigue caused by too much current. We have combined the mass spectrometer with a vacuum system which is connected directly to the spectrometer and is very useful for sample preparation, for synthesizing samples, and for loading them into the capillary gas chromatographic column.

Recently we have combined paper chromatography and mass spectrometry in a way which looks quite promising. In the future we hope to do for paper chromatography what the mass spectrometer has done for gas chromatography, thus increasing its potential greatly by combining the two techniques. A rough sketch of a gas chromatograph with a mass spectrometer attached is shown in Figure 2-1. The spiral is the 300-ft column. Conventionally, such a small-diameter column can accept only $10^{-3}$ to $10^{-2}$ $\mu$l of a sample without overloading, which means that if the sample is in a solvent only very small samples can be used and, therefore, exceedingly sensitive detectors are required. Loading is normally done by flowing helium through the whole system to carry the organic materials—by squirting in the small sample, bypassing 99.9 per cent of it, and allowing the rest to flow in the column. Normal procedure also requires use of a helium separator before the sample passes into the mass spectrometer; otherwise, the pressure is built up so high that the mass spectrometer will not operate.

In our system we have replaced the inlet with a vacuum system so that we aliquot the gases at various temperatures, then put them on the column by vacuum distillation. We put a little U-tube in the column with a cold trap on it and collect our sample. We then remove the cold trap and develop the

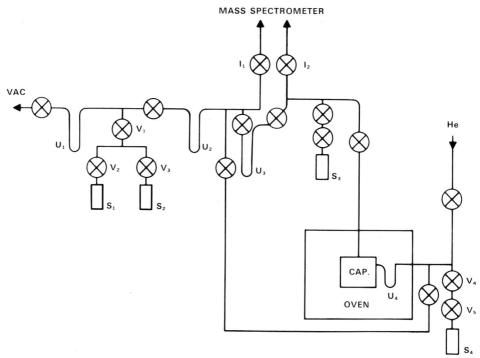

**Fig. 2-2. The vacuum system used for sample manipulation.**

column with helium. We find that the Bendix has a fast enough pumping system; large tubes without any constrictions have very fast pumping speeds. We do not need to use a helium separator. With a $300 column and some cheap plumbing we have made a very powerful analytical tool out of this system. The mass spectrometer is the detector for the gas chromatograph and is also a means by which we can get mass spectra at any time while the sample is coming off the capillary column. The column we use is equivalent to approximately a hundred thousand theoretical plates, which means a theoretical exchange every millimeter or less. Such performance yields some rather remarkable separations.

The sample handling system is illustrated in Figure 2-2. It is a collection of valves and ¼-in. copper tubing. It is almost entirely a metal system. We put a meteorite sample in the sample tube, evacuate the system, and then start warming it. We have a liquid nitrogen trap on the U-tube. We warm the sample at various temperatures and let it remain overnight. The analysis is run the next day. We "crack" this valve open to let the U-tube warm up conducting the sample back to point $S_3$. This way we can look at some rather volatile fractions and by this very simple system frequently get rather good separations. At the outset we can learn a great deal about a sample without putting it through a capillary column. Another experiment we have done is to introduce a synthetic sample at point $S_1$ where we start heating a mixture of gases. At known times we take an aliquot between $V_1$, $V_2$, $V_3$. We pump off volatile non-condensibles through this U-tube, look at it in the machine, and observe the products. Alternatively, we can condense it back to point $S_2$, put it on the end of the capillary column, and flush out as described above. We have provided for a temperature bath on the sample. We take an aliquot between the two valves $V_4$ and $V_5$, close valve $V_5$, close off the sample, load the capillary column and develop it. Two advantages of this system are that it is a closed system and that various sections can be isolated. Thus we can minimize the contamination hazard which is horrendous in work with very tiny amounts of organic material, because the whole world is organic: fingerprints, shaving lotion, cigarettes, or what have you. A third advantage of loading the column by vacuum is that a larger sample can be loaded without losing resolution. We load a sample and let it develop on the column to allow it to band. We can get the more volatiles to appear at the far end of the band and the less volatiles at the near end. Even though the peaks may be less sharply defined, we can load a large amount of material and look at extremely minor components.

Gas chromatograms of nonane-xylene mixtures are shown in Figure 2-3. This is a check-out system to illustrate the capabilities of the machine. It also illustrates very well the value of a combination of a mass spectrometer and a gas chromatograph. One-ml aliquots were taken from the sample in Figure 2-2 at dry ice temperature. Three chromatograms are represented, each one run at different conditions. Ethyl benzene was a contaminant in the commercial xylene. At the top of the figure we see that we resolved ethyl benzene, two xylenes, and nonane. A very slight break shows a very slight resolution between meta and para xylene. For all three chromatograms the mass spectrometer was tuned in such a way that we looked at all the masses between 50 and 150 by means of the blanking generator. Everything else was blanked out. The ordinate is the response of the multiplier in this mass region. As we developed the column, we obtained a series of mass spectra across each of the peaks. We could take a dozen mass spectra across the peaks if necessary. In the figure, with different conditions, we see a better resolution of para xylene and meta xylene but the nonane and ortho-xylene are now coalesced into one peak. In the next run, under isothermal conditions of 45° C. these were resolved again, but the order was reversed. The aromatics have shifted relative to the aliphatics. The mass spectra of all three of the

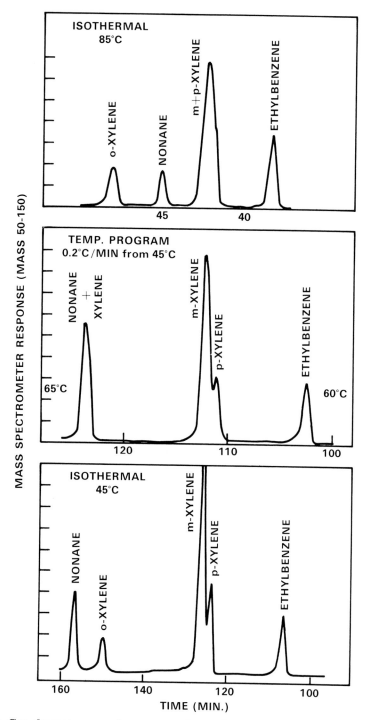

Fig. 2-3. Gas chromatograms of a nonane-xylene mixture. This was used to check out the entire system.

xylenes are essentially identical, so it would be impossible to determine that three xylenes were present without chromatographic evidence to show that the peaks were actually separated.

The presence of a xylene mass spectrum coupled with the three chromatographic peaks is a conclusive evidence for the ortho, meta, and para forms. Figure 2-4 shows a resolution of that peak where the nonane and ortho-xylene appear as a single symmetric peak. We made a dozen scans, one at the beginning, one at the middle, one at the end as shown in the figure. The parent peak of nonane is very evident in the top spectrum; it is essentially all nonane. The middle one is a mixture of nonane and xylene. The lower scan is essentially only xylene with masses 106 and 91. This illustrates the technique and the power of the combination of a mass spectrometer with a gas chromatograph.

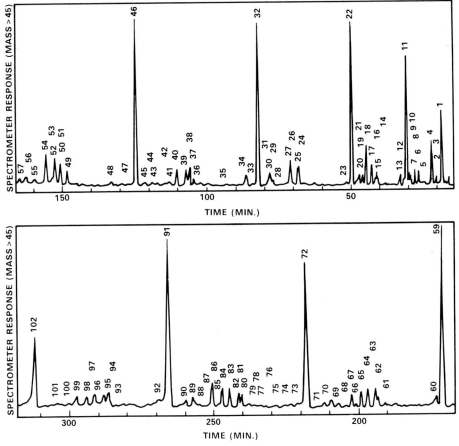

Fig. 2-4. Three mass spectra across composite nonane o-xylene peak showing resolution by mass spectrometer.

METEORITE ANALYSIS

Some synthetic experiments shed light on the organic matter that we found in the meteorite. Two hypotheses have been proposed to explain the origin of these organic materials. One of them proposes a biogenic origin. Perhaps the organic matter was from a planet that disintegrated and which had contained life, or it came from the moon. Nagy, Meinschein, and Urey are among those who have held this position in the past. Another hypothesis is that the organic matter was perhaps formed by a so-called Miller-Urey type of reaction, where high energy electric discharge or ultraviolet light is used to excite molecules and to produce organic compounds from simple gases.

In 1965, we published a paper with Anders and Hayatsu in *Science* based on an analysis of some volatile material from the chondrites. Being amateurs, we first ground up the chondrites and then warmed them up from room temperature. We detected the volatile materials that came out. The previous conventional experiments had been done with solvent extractions in which the very volatile components would all be lost by the time the experimenter had an opportunity to search for them. We were surprised to see some highly volatile substances such as methane, hydrogen, and carbon monoxide. When we heated the sample to higher temperature we detected benzene, toluene, and a variety of simple aromatic compounds. The principle inorganic compounds found were aromatics and methane. The results suggested to us that some sort of a reaction approaching equilibrium produced these compounds. This suggestion met with some criticism, especially from Urey, who says that he believes the organic matter in meteorites could have been formed only by high energy reactions or by living organisms. Since then, we've done approximately one hundred experiments, a large fraction of which were synthetic experiments to determine the conditions under which the organic matter was formed. The results leave little doubt about the reasonableness of our point of view.

The substantiating evidence is shown in a series of figures. Figure 2-5 shows the results from the high energy reactions and the type of distribution obtained from the so-called Fischer-Tropsch type of reaction, which is a reaction between carbon monoxide and hydrogen with an iron type catalyst. A high energy reaction in a spark discharge of methane produces many isomers. There is such a tremendous number of them that on the gas chromatogram it looks just like a continuum, whereas the Fischer-Tropsch type is a very selective process. Each of the large spikes is the normal alkane, and there is a less abundant group in between. The Fischer-Tropsch type is much more selective and gives a much more limited number of compounds. We chose to work with carbon monoxide and hydrogen as simple gases because they are believed to have been present in the earlier solar system in this form, especially when it was at high temperatuers. Also, we found them in the meteorites as trapped gases.

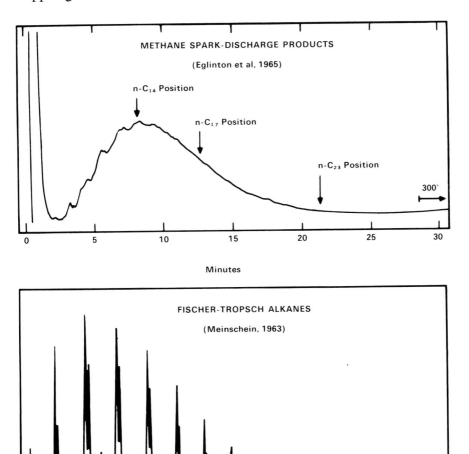

Fig. 2-5. Comparison of results from a high energy reaction and from the Fischer-Tropsch reaction using CO and $H_2$ with an iron catalyst.

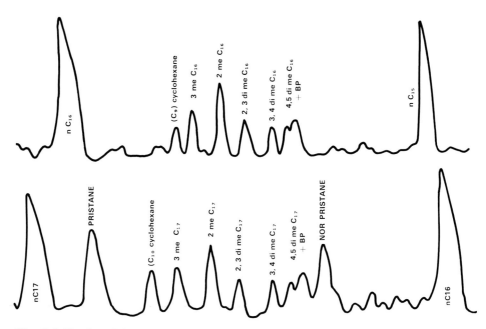

**Fig. 2-6. Results of heating CO and D₂ in the presence of a Canyon Diablo sample as catalyst.**

We have substituted deuterium for hydrogen so that we can eliminate any possibility of contamination in our synthetic experiments. All our compounds are completely deuterated so that all show a preponderance of even masses. Thus, it is very easy to pick out a chance contaminant. We found that we needed a catalyst in order to get any appreciable reaction in a reasonable length of time. We used as a catalyst an iron meteorite, Canyon Diablo. We baked it to eliminate any organic matter and then put it through an oxidation-reduction cycle. We oxidized it to oxide and then reduced it, with deuterium to metal, and used this as a catalyst.

Figure 2-6 shows a summary of the results of heating a mixture of carbon monoxide and deuterium in the presence of Canyon Diablo starting at room temperature and sampling it in the mass spectrometer as the temperature was increased. What happens if this process is done with a 50-50 mixture of carbon monoxide and deuterium and a catalyst? One sees methane building up with one carbon atom, then ethane, propane, etc., and one may go all the way up to 20 or 30 carbon atoms in a chain. As the ratio of carbon monoxide to hydrogen is changed, the ultimate chain length is changed. The higher the ratio of carbon monoxide to hydrogen the longer the chain length. In one case, we observed a chain of 28 carbon atoms with a ratio of hydrogen to carbon monoxide of one. With a ratio of 250, the longest aliphatic chain we could build up was 12 carbon atoms. Some substances such as aluminum oxide favor the production of olefins. Even without the presence of aluminum oxide, some olefins can always be seen even when the ratio of hydrogen to carbon monoxide is very high. This is a clue to the formations of some of the hydrocarbons produced by this type of reaction.

When heating is continued cyclo-paraffins form and then benzene derivatives appear. As sampling continues the long chains can be seen folding over on themselves and forming benzene derivatives with very long side chains. With further heating, the side chains form polynuclear aromatics or the side chains break off and toluene builds up. At still higher temperature the ratio of benzene to toluene increases. When it gets quite hot, a great abundance of ethylene is noticed, and some of the benzene may be formed by a direct polymerization of benzene. Finally, by heating all the way to 900°C. one can see essentially only aromatics and methane, water and carbon dioxide. Figure 2-7 is a chromatogram of a single loading on the capillary column showing a sample taken at one stage of one of these heating experiments. We heated a 50-50 mixture of carbon monoxide and deuterium for several days, then sampled. This sample was taken after heating to 300°C. The gas aliquot contained compounds from 5 to 14 carbon atoms. The big spikes are normal alkanes. In the lighter region there are many aromatics such as toluene, benzene, and xylene; however, with higher and higher masses, a repetitive pattern of alkanes shows up. The pattern consists of prominent normal alkanes with less abundant but well-defined branched alkanes between each adjacent pair of normal alkanes.

It is important to note the selectivity of this method of producing hydrocarbons. They are relatively few. The predominant ones are the normal alkanes. A similar fact is observed in biogenic materials with a predominance of normal alkanes and a small amount of branched alkanes. Incidentally, in this single run we identified one hundred different compounds. (The analysis required quite a bit of time.)

Figure 2-8 shows the same repetitive pattern. This was another run at a lower temperature. It was more nearly purely aliphatic, essentially no aromatics. In this run the repetitive pattern repeated itself from 9 carbon atoms up through 17 carbon atoms, which was as far as we built up the chain. We looked at representative samples from each of these characteristic peaks. The ordinate of the chromatogram is an integrated current which is proportional to the number of mass spectral peaks in a broad region taken in the mass spectrometer. We took representatives from each characteristic peak for comparison. It turns out that the first peak is always a 3-methyl alkane, the second

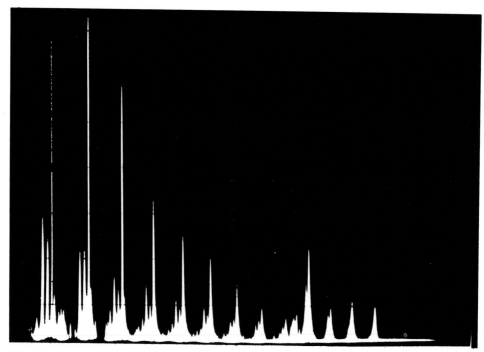

**Fig. 2-7. Chromatogram of a single loading of the 300-ft capillary column at one stage of the heating experiments.**

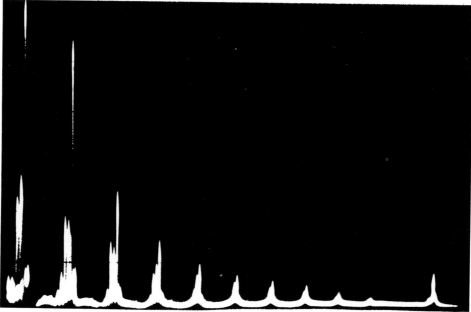

**Fig. 2-8. Example of repetitive pattern of aliphatics from sample at lower temperature. There are essentially no aromatics.**

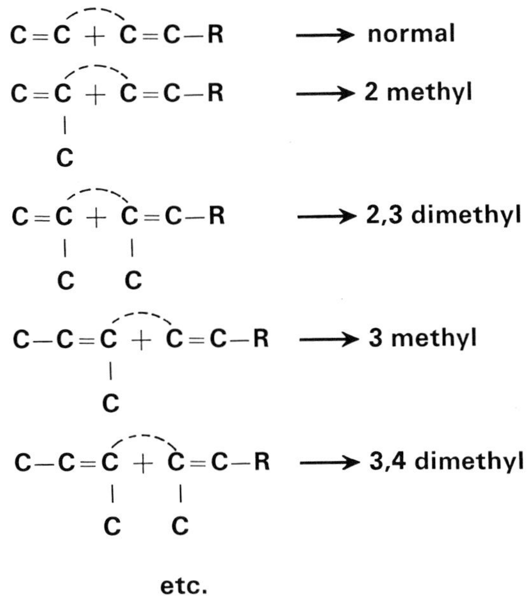

$$C = C + C = C - R \longrightarrow \text{normal}$$

$$\begin{array}{c} C = C + C = C - R \\ | \\ C \end{array} \longrightarrow \text{2 methyl}$$

$$\begin{array}{c} C = C + C = C - R \\ | \quad | \\ C \quad C \end{array} \longrightarrow \text{2,3 dimethyl}$$

$$\begin{array}{c} C - C = C + C = C - R \\ | \\ C \end{array} \longrightarrow \text{3 methyl}$$

$$\begin{array}{c} C - C = C + C = C - R \\ | \quad | \\ C \quad C \end{array} \longrightarrow \text{3,4 dimethyl}$$

**Fig. 2-9. Condensation of olefins.**

etc.

one a 2-methyl alkane, the third one 2,3-dimethyl, then 3,4-dimethyl, and the one that's starting to build up apparently is 4,5-dimethyl, as shown in Figure 2-9.

To show how we reach this conclusion, I shall describe the mass spectra from organic compounds. Different classes of organic compounds have very characteristic fragmentation patterns. A straight chain aliphatic will break up into ionized fragments consisting of 1-2-3-4-5-carbon atoms, all the way up to the parent ion, with the maximum occurring usually at 3 and 4 carbon atoms for a long, straight chain compound. When a branch is present, there will be a spike at the branch. A Polaroid photograph of the oscilloscope screen is shown in Figure 2-10. This is hexadecane. The parent ion has 16 carbon atoms. Loss of one methyl group is highly improbable; it cannot be seen on this spectrum. Figure 2-11 is a fast scan of the same sample of hexadecane. The next one is an iso-hexadecane, that is, it has a methyl group on the second carbon atom. We no longer have that first methyl group missing; we lose a methyl group because of the branching and we have a prominent peak with three carbon fragments lost also. (The spurious peak is mercury from the diffusion pump system.)

Using this technique we deduced the structure of the branched aliphatics between the normals. A naive representation of how these branched and straight chain links form is shown in Figure 2-12. When I first looked at this I was struck by the repetitive pattern time after time with always about the same

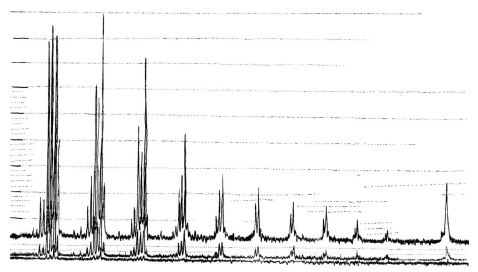

**Fig. 2-10.** Polaroid photograph of the oscilloscope display of the time-of-flight mass spectrometer output for hexadecane.

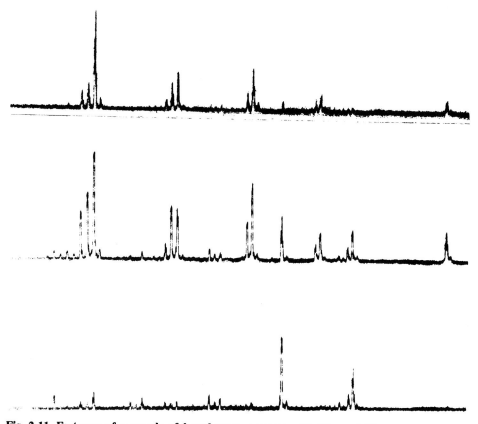

**Fig. 2-11.** Fast scan of a sample of hexadecane; compare with Figure 2-10.

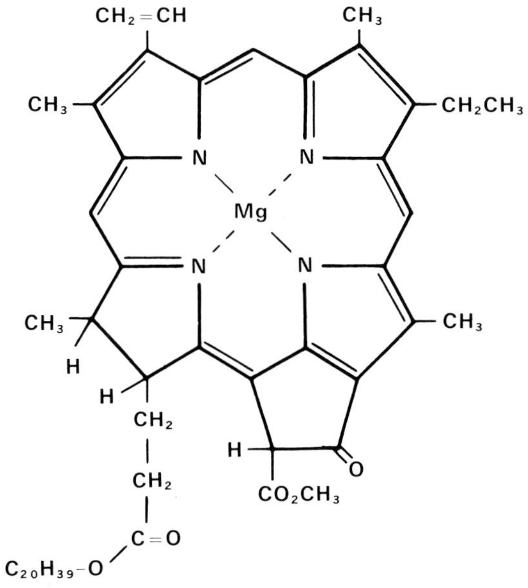

Chlorophyll

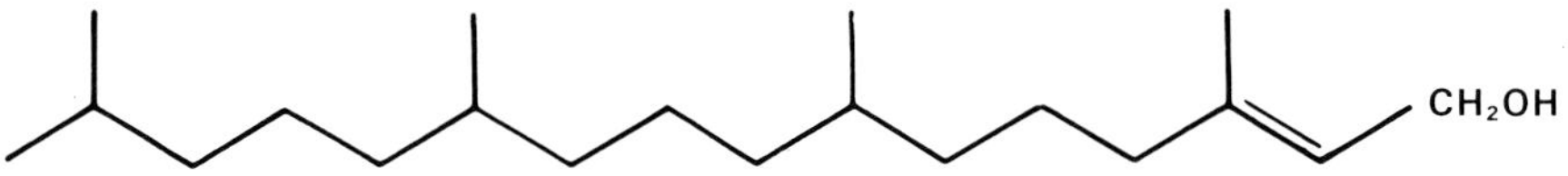

Phytol = $C_{20}H_{39}OH$

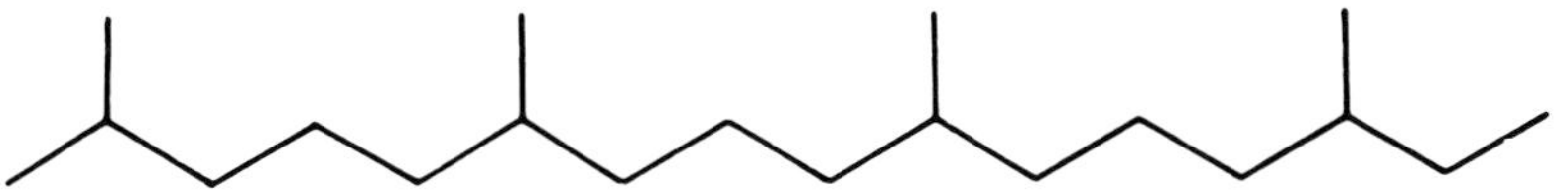

Phytane = $C_{20}H_{42}$

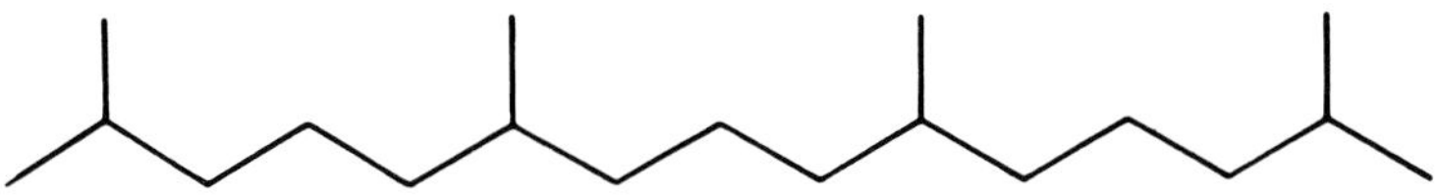

Pristane = $C_{19}H_{40}$

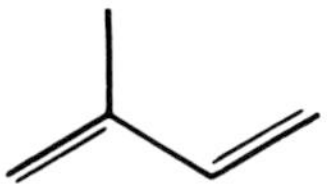

Isoprene = $C_5H_8$

**Fig. 2-12. A representation of the formation of the straight chains and the branched aliphatics.**

ratio of normals to branched. It suggested to me that there must be a common mode of formation for all of them. This may be simply a way in which the olefin could combine and form these compounds without going into any details of mechanism, combined with the fact that light olefins are always abundant whenever synthetic experiments are done. Figure 2-13 is an example of an experiment representing the ultimate limit obtained when aliphatic compounds are heated. This one actually was from heating methane and shows various polynuclear aromatics. The same sort of results occur by building up the chains and then continuing to heat.

**Figure 2-13**

**AS TEMPERATURE OF MIXTURE OF CO + H$_2$ + CATALYST IS INCREASED**

1. ALIPHATIC CHAINS BUILD UP C$_1$, C$_2$ ——————————C$_{20}$ — —
AS RATIO OF CO TO H$_2$ INCREASES CHAIN LENGTH INCREASES.
AL$_2$O$_3$ FAVORS OLEFINS — — ALWAYS SOME PRESENT.
2. CYCLOPARAFFINS AND BENZENE DERIVATIVES FORM.
3. TOLUENE BUILDS UP.
4. POLYNUCLEAR AROMATICS FORM.
5. RATIO OF BENZENE TO TOLUENE INCREASES.
6. FINALLY ONLY AROMATICS AND METHANE.

It is of general interest to tabulate a few of the aromatic organic compounds that have been isolated from meteorites. These are shown in Figure 2-14.

ORGANIC COMPOUNDS IN METEORITES

A comparison of our synthetic experiments with the organic compounds which we have found in meteorites is historically the reverse of the way we investigated the problem but perhaps a bit more logical for understanding our technique. The principal meteorites used were Orgueil meteorites and Murray. A variety of experiments were designed in an attempt to eliminate all possible contaminants. I sometimes feel that the extreme precautions we take in the laboratory are ridiculous when we don't know whether the sample was carried in somebody's coat pocket on the way to the museum.

We ground the meteorite and distilled it at various temperatures, for example, 100°C., 150°C., or 200°C. At about 200°C. we started to synthesize compounds such as toluene derivatives which would appear in abundance; so we would usually keep below that temperature. We extracted the products with a group of solvents. We used iso-pentane, which is a low-boiling solvent, in an attempt to preserve the more volatile materials that would be lost with a higher boiling solvent when it was attempted to remove the solvent. We tried

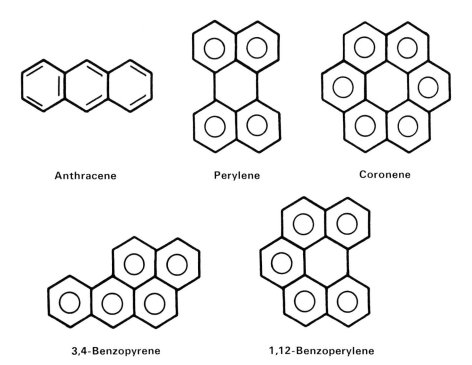

Fig. 2-14. Representative aromatic organic compounds isolated from meteorites.

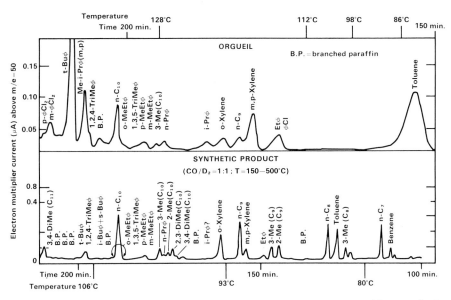

Fig. 2-15. An Orgueil meteorite fraction distilled at 150° C. compared with a synthetic mixture.

benzene, benzene-methanol, and heptane. There was in general a great deal of consistency among the compounds observed in all these various modes of isolation. An inconsistent result when found was attributed to a contaminant. There seemed to be a prevalence of contaminants used for mothballs; for example, dichlorobenzene, naphthalene, or camphor.

Figure 2-15 shows a fraction of the Orgueil meteorites distilled at 150° C., compared with a synthetic mixture we gradually built up and then heated for a brief time at 500°C. This procedure synthesized a whole host of aromatic compounds such as benzene and the xylenes. We can see meta- and para-xylene resolved here. In the meteorites we have a large toluene peak. Here's a chlorobenzene which everybody finds in meteorites. We sometimes find it as a large sport, and we wonder whether it could have been a contaminant at some time or other. In these two runs there are sixteen compounds that are common to both. The aromatics are prominent in both cases. The light aliphatics are completely missing in the Orgueil; the first one seen is the C-9. Quantitatively there is not a one-to-one correspondence, but qualitatively they look very similar. They are not on the same time scale because at that time there was no rigid temperature program; the oven was just turned up every once in a while, so that it was difficult to put them on the same basis without a great deal of replotting. The substituted benzenes that are very prevalent have side chains of from 1 to 4 carbon atoms. This seems to be characteristic of all the samples of all the meteorites looked at; it is also very characteristic of what is found in the synthetic experiments with carbon monoxide and deuterium.

The results of the analysis of Murray under the same conditions are shown in Figure 2-16. The sample was distilled at 150° C.; all the normal alkanes and all the substituted benzenes are identical in the two samples. We particularly wanted to discover whether we'd see this repetitive pattern of the branch alkanes; however, there are so many aromatics in both of these experiments that it's difficult to pick them out, although they can be identified in the synthetic runs. Table 7 compares some of the abundances between the meteorites and the synthetics. This illustrates the general agreement among the synthetic experiments. So many things can change the distribution that, even though we have not accidentally obtained a complete match of all details, we are convinced that any hydrocarbon which can be shown to be indigenous to the meteorites under the proper chosen conditions can be synthesized from carbon monoxide and hydrogen. A higher region is shown in Table 8. Here the aromatics begin to disappear, and resemblance between the meteorites and the synthetic products now becomes much more prevalent. A first glance at only

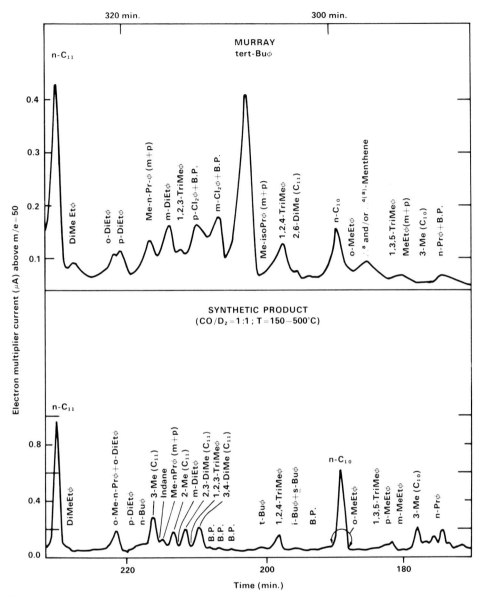

**Fig. 2-16. Analysis of the Murray meteorite under conditions identical to those in the CO-D₂ synthetic experiment.**

the chromatographic data, without any mass spectrum analysis, would suggest they're identical. However, a closer examination including a comparison with the mass spectrum reveals that the first two peaks are identical, and the other ones are again branched carbons, but they are not the *same* ones. We have not yet deduced what they are, but they are clearly not the same ones. This

**Table 7. Comparison of Abundances in the Meteorites and in the Synthetics.**

| | Hydrocarbon Distribution, $C_6$ - $C_{10}$ | | |
| --- | --- | --- | --- |
| | Orgueil % | Murray % | Synthetic % |
| Normal paraffins | 17 | 10 | 44 |
| Branched paraffins | 0.5 | 10 | 22 |
| Olefins | 0 | 1 | 0 |
| Aromatics | 82 | 79 | 34 |

**Table 8. Continuation of Table 7 into a Higher Mass Region.**

| | Aliphatic Hydrocarbons, $C_{11}$ - $C_{18}$ | | |
| --- | --- | --- | --- |
| | Orgueil % | Murray % | Synthetic % |
| Normal paraffins | 58 | 57 | 52 |
| Branched paraffins | 35 | 32 | 41 |
| Olefins | 7 | 10 | 7 |

might have been the result of isomerization over long periods or of the effects of cosmic rays, if they had a common origin. I shall return to this problem.

We completed an experiment just a week ago that shows the correspondence between the different types of compounds, where the synthetic closely resembles the compounds found in shale. A sample of shale oil, from a billion-year-old Nonesuch shale from Michigan, was given to us by Warren Meinschein, professor at Indiana University. In comparing it with the synthetic hydrocarbon fraction, these show a marked one-to-one correspondence between the shale and the synthetic. The extra peak in the shale is a cyclo-hexyl derivative, which can also be seen occasionally in synthetic samples. There is the 3-methyl, just as in the synthetic. There is the 2-methyl, and the 2,3-dimethyl, the 3,4-dimethyl, and the 4,5-dimethyl. In addition to those compounds, there are a couple of compounds in shale which have been used in biogenic tracers in the past. This one is pristane; it is a branched hydrocarbon with 19 carbon atoms. It is 2,6,10,14- tetramethyl pentadecane. In other words, it is a 15-carbon atom chain with 4 symmetrically spaced methyl groups. This is known to be a biogenic material produced from living materials. But there is also a suggestion that it can be synthesized. Pristane has actually been synthesized from isoprene, experimentally with a specific catalyst. We have found a large group of related compounds—the so-called isoprenoid compounds—which have isoprene as the basic unit, and which can be thought of as polymers of isoprene. The C-18 isoprenoid is also present in the shale. There are some more highly branched compounds, and olefins, very analogous to synthetic mixtures.

The questions are: "Since this shale oil is thought to be of biogenic origin and there is some optical activity observed in it, does the biogenic material have the same distribution pattern as does the synthetic, or is this mostly synthetic contaminated with biogenic materials; or are these materials not biogenic; or were they synthesized in some slightly different condition?" Referring back to Figure 2-12, the origin of phytane, a C-20 isoprenoid, is chlorophyll. The side group is the phyto group. Here is pristane, where all the corners are carbon atoms, showing the symmetrical spacing. Isoprene is a di-olefin with 5 carbon atoms including a methyl group branch. We have never seen this in a meteorite, and we have never seen it in a synthetic, but we've seen a close relative both in the meteorite and in the synthetic, isopentene, with only one double bond. The observation suggested to us that maybe we do synthesize the isoprenes by condensation of isopentene. Some of the isoprenoids which we synthesized in the synthetic mixture are shown in the gas chromatogram. This

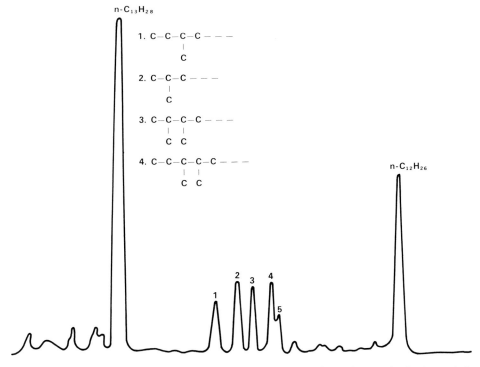

Fig. 2-17. Synthetic deuterated isoprenoid compounds. A prominent feature is the loss of 8 carbon atoms.

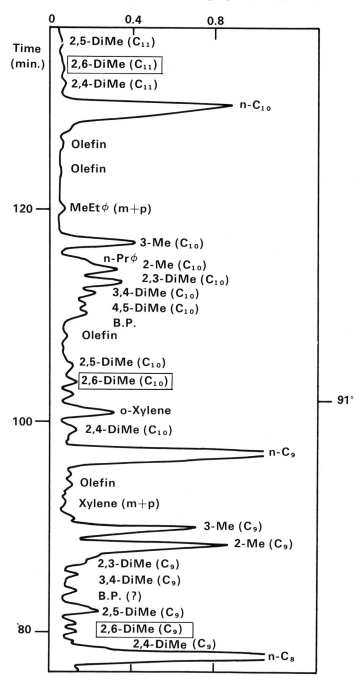

**Electron multiplier current ($\mu$A) above m/e=50**

Fig. 2-18. A 10:1 mixture of $D_2$ and CO heated to 200° C.

is a 10-to-1 mixture heated to 200°C. The mass spectra of two $C_8$ compounds may appear very similar with one the iso and one the 2,6-isoprenoid. How-

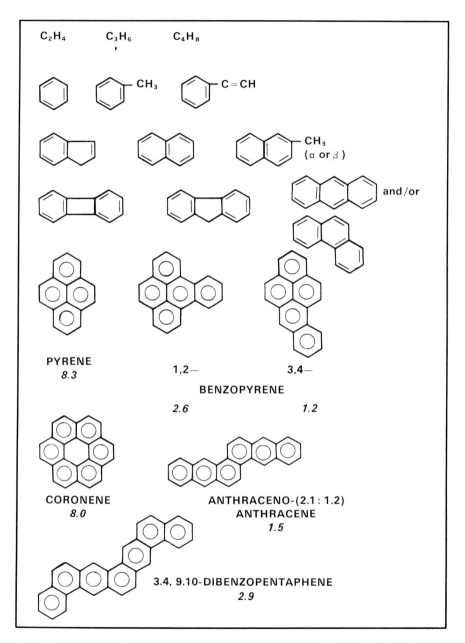

Fig. 2-19. Representative organic compounds synthesized in $CH_4$ or $CH_4$-CO-$CO_2$ mixture at 900° C.

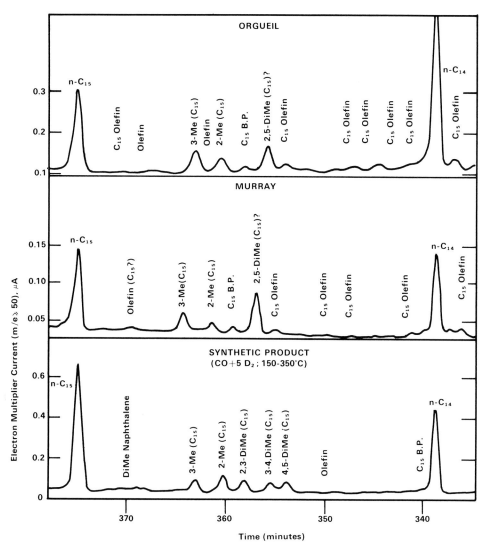

**Fig. 2-20. Comparison of the Orgueil, the Murray, and the synthetic.**

ever, one can tell the difference between the two because the more highly branched one is more volatile and appears sooner in the chromatogram. We observed several such pattern similarities which we assign as isoprenoids, because the volatility fits and the mass spectra fit. If this were some other closely related compound giving the same mass spectra we wouldn't be able to differentiate. The mass spectra of some synthetic isoprenoid compounds are shown in Figure 2-17, They have a characteristic mass spectrum in that the loss of 8 carbon atoms is always prominent. The mass of the deuterated compound is 130. We have deduced from both their positions in the

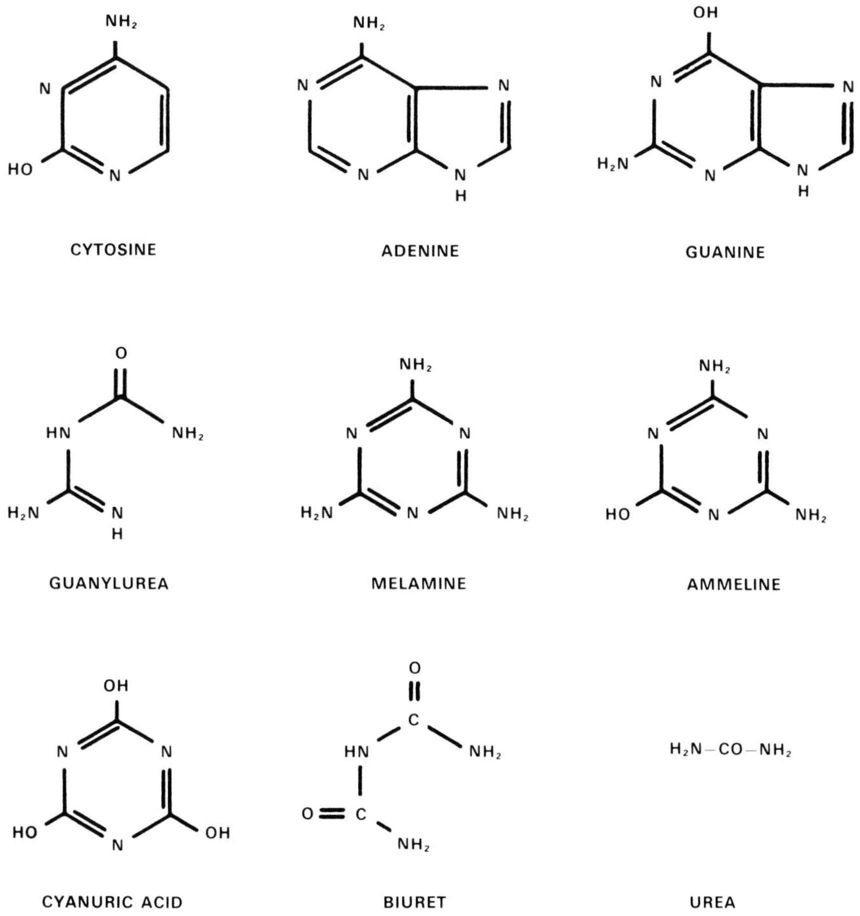

**Fig. 2-21. Structure of compounds formed in ammoniated synthetic experiment.**

chromatograph and the mass spectrum analysis that these peaks are the C-10, 11, 12, 13 and 14 isoprenoids which we have synthesized. Other examples of synthetic experiments are shown in Figures 2-18 and 2-19. There is a small peak that we identified as isoprenoid in the mass spectrum of the Murray meteorite. We also saw a couple of others in the Murray (Figure 2-20). But we did not see any pristane or phytane. We have not seen pristane or phytane in any meteorite, although it has been reported by Oro at Houston. He had a large number of organic carbonaceous chondrites; in samples of the same meteorite he sometimes found pristane, sometimes did not. This suggests to me that there may have been accidental contamination, sometime in the history of the meteorites. Thus it is not clear whether pristane and phytane, the so-called biogenic indicators, are really present in meteorites.

**Table 9. Products of Synthetic Experiment in Which Ammonia Was Added to the Standard CO-H₂ System.**

Organic Compounds from CO, H₂, and NH₃ (1 : 5 : 0.2; 200° C; 6 hr; iron meteorite catalyst)

| | | |
|---|---|---|
| $CH_3NH_2$ | $CH_4$ | Benzene |
| $(CH_3)_2NH$ | $C_2H_4$ | Toluene |
| $(CH_3)_3N$ | $C_2H_6$ | Xylene |
| HCN | $C_3H_6$ | |
| $CH_3CN$ | $C_4H_8$ | |
| $C_2H_5CN$ | 2-Me-propene | |
| i-$C_3H_7CN$ | $C_5H_{10}$ | |
| $C_4H_8(CN)_2$ | 1-Me-pentene | |
| | $C_6H_{12}$ | |
| | 5-Me-hexene | |

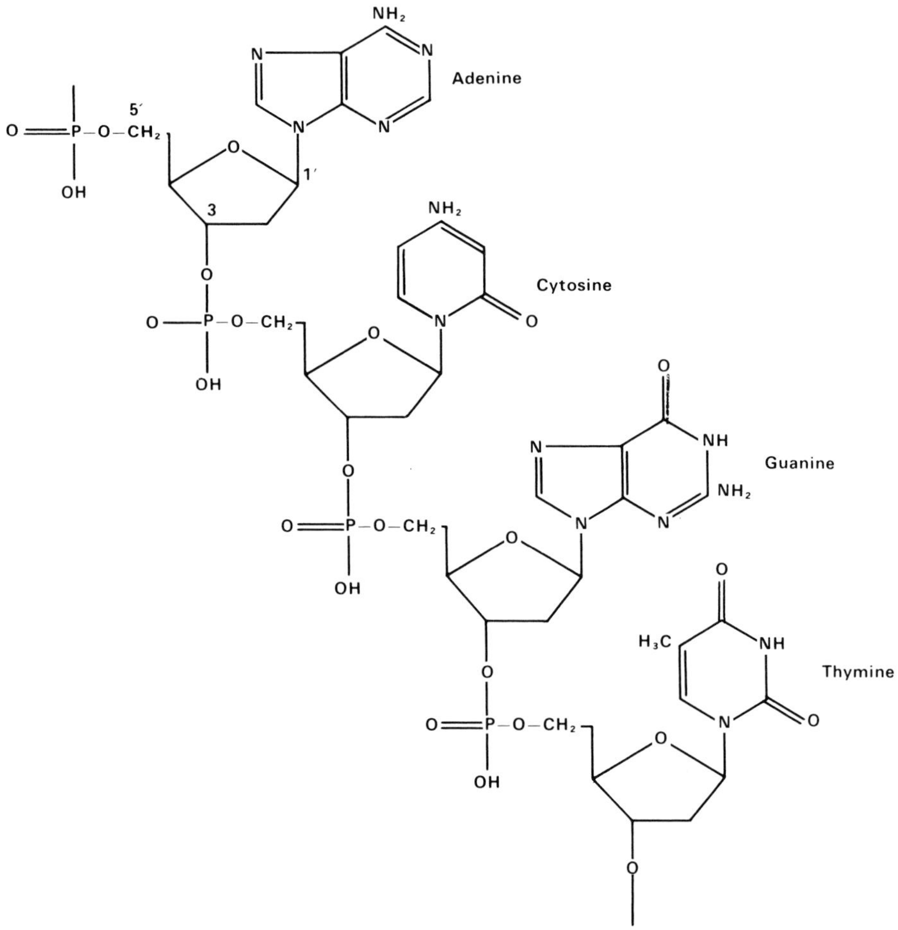

Fig. 2-22. The DNA molecule.

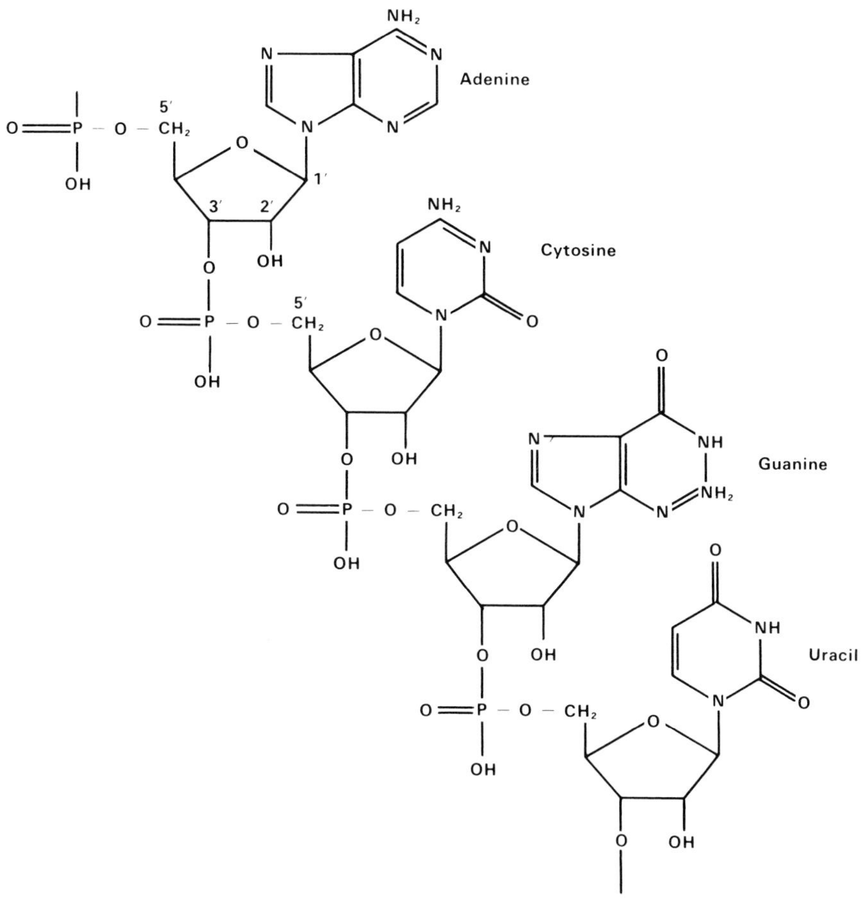

**Fig. 2-23. The RNA molecule.**

NITROGEN COMPOUNDS

I now shall report on some work we have done with nitrogen compounds. Quite a number of nitrogen compounds have been reported in meteorites. To investigate this we added ammonia to our simple system of carbon monoxide and hydrogen. (Chronologically this was before we changed over to deuterium.) This mixture was heated in the usual manner. Some of the compounds identified by mass spectrum analysis are shown in Table 9 and Figure 2-21.

These are some of the organic compounds that were synthesized. Most of these identifications were made by Hayatsu, by a very extensive series of chromatographic techniques, column techniques, and paper chromatography. Of those first six all were found in the Orgueil meteorite except cytosine. The interesting feature of these data is that the synthesized substances are important biologically in DNA and RNA. In this picture of DNA (Figure 2-22), one can find adenine, cytosine, guanine, thymine. Figure 2-23 is RNA— again very similar structure with uracil being substituted for one of the others.

**Figure 2-24. Summary of relevant evidence.**

1. NOT NECESSARY TO ASSUME BIOGENIC ORIGIN OF O.M. IN METEORITES.
2. SPECIFICITY OF F.T. TYPE REACTION CASTS DOUBT ON VALIDITY OF USE OF PRISTANE AND PHYTANE AS BIOLOGICAL INDICATORS.
3. MILLER-UREY TYPE REACTIONS CANNOT ACCOUNT FOR O.M. DISTRIBUTION IN METEORITES.
4. BIOLOGICALLY SIGNIFICANT COMPOUNDS SEEM TO BE EASY TO MAKE.
5. OPTICAL ACTIVITY PROBABLY ONLY RELIABLE INDICATOR FOR BIOGENIC ORIGIN.

CONCLUSIONS

A summary of the evidence as shown in Figure 2-24 suggests to me that it is not necessary to assume the biogenic origin of organic matter in meteorites to account for what we find there. It is clear they could have been produced from carbon monoxide and hydrogen perhaps in the early solar system before it condensed into planets, or maybe captured in the bodies at a later time and heated subsequently. The specificity of this type of reaction, I feel, casts some doubt on the validity of the use of pristane and phytane as biological indicators, especially since a large group of related compounds have been made. We believe that occasionally we have made even pristane, but not in very great abundance. Going from the dimeric species to the trimeric, the probability drops off rather rapidly. We feel that the Miller-Urey type of reaction could not account for it unless it was subsequently modified by biological activity, because the Miller-Urey type gives a continuum of a very large number of isomers and does not at all approach the distribution found in the meteorite. It seems that in general the compounds that are biologically significant appear to be the ones that are easy to make. If life originated after a big storehouse of organic materials appeared, it is logical that they would feed on only what they had available. Those substances that were easily made would be the food that the biological system tapped as it was continuing to reproduce. It seems to me that optical activity is probably the only reliable indicator of a biogenic origin. If in some other system there was life without optical activity we would be forced to take an entirely different approach.

DISCUSSION

(Q) Optical activity is normally an indication of a life process. Have you isolated sufficient material that would be either the L or D variety so that you might be able to say that it did or did not exist in the meteorite?

(A) Studier: A man by the name of Nagy working with Urey on the West Coast reported optical activity in the Orgueil meteorite, but Hayatsu tried to repeat this in vain. He did a great lot of work on it and concluded that it was a very, very tiny effect. He concluded that the effect was an artifact. I think it is probably generally conceded that there is no good evidence for optical activity in the compounds from meteorites. In meteorites that were obviously contaminated due to other indicators, optical activity has been seen, but I think nobody believes that the presence of optical activity in carbonaceous chondrites has been proved.

(Q) When you take typical molecules, such as the iron in meteorites, and tag them with oxygen or other atoms, you get a very complex series of reactions. Have you considered *all* those possible reactions and their use as catalysts?

(A) Studier: Certainly the build-up of the aliphatic chains are catalyzed. No catalysts were used in the polynuclear aromatics that I showed on the charts; there was just straightforward thermal heating. If you take any of the aliphatic hydrocarbons, I don't care what they are, and heat them to a high temperature, you'll form aromatics. If you continue heating, you'll probably eventually form carbons, but in the interim you always find the "junky mess" that is attributable to the high molecular weight aromatic compounds. I don't think you need a catalyst for that, because by then you have reached a much higher temperature. Incidentally, as a sidelight, we went to an active cobalt catalyst, which was specially prepared to be very active, and did this same series of experiments. All the branch compounds dropped down by about a factor of 20. At first I thought we'd completely eliminated them, but then loading a column heavier I saw that all were still present but that the ratio to the alkanes was done by a factor of 20, so instead of 20 per cent they were only 1 per cent or less. This shows how just slight changes in conditions can completely change the whole distribution pattern. I think the valid approach is to determine what materials are found in the meteorites and then devise experiments to see how these might have been formed. On the basis of what we've observed, I feel that the long chain hydrocarbons were probably built up first and then later the material was reheated to form, partially, aromatics such as those found in the light region which are predominantly aromatics. As you go to the heavier ones you still see them. Maybe the heating was not uniform. Maybe it was similar to what we've been doing in the laboratory. Our experiments usually have been done with a long tube where we're just heating the bottom of it, and then as things heat up the species can distill off and their re-entry into the hot zone is limited. So you preserve them for a while and can look at them.

(Q) Have you done any experiments with spark discharges?

(A) Studier: We have not. There are groups at NASA, the Ames Research Center near San Francisco, who have done a great deal of it. Of course, these are the classic Miller-Urey experiments. Many people are doing this kind of experiment. It's true that such discharges can make anything under the sun; you can even make amino acids if you wish. We haven't looked for amino acids in meteorites yet because they're such a tremendous contamination hazard. We've just now gone over to a deuterated ammonia so that we can spot contaminants, and we will look at the amino acids at a later time.

(Q) I want to check a guess. You had a column made up which was 0.01 in. through and 300 ft long. This was apparently essential to the whole argument. I understand that you somehow diffused in organic materials. Was it a liquid-filled column or a gas-filled column?

(A) Studier: The column has just a coating of apiazon L on the inside. The whole thing is evacuated at the beginning. Then we distill off a sample by vacuum distillation, by heating, and by flushing it on; then the temperature of the whole thing is gradually increased while the helium is flowing. The different compounds separate into bands as they go along the column.

(Q) Because they diffuse in different ways. What is the effect of the helium?

(A) Studier: When you're using apiazon L with aliphatic hydrocarbons, it is almost exactly a measure of vapor pressure.

(Q) OK; does the position of the various materials depend upon their difference in vapor pressure?

(A) Studier: You take advantage of the very slight difference in the vapor pressure. It really is a distribution between the organic compound dissolving in the materials of the wall, and coming out again—the material dissolving and coming out again. If you're working only with aliphatic compounds (this material we have here is an aliphatic) it's almost exactly proportional to the vapor pressure. For example, methane will come through first, then ethane, then propane, etc.

(Q) In your mass spectrometer does the time of arrival somehow depend upon the mass?

(A) Studier: It's the time of arrival. I didn't go into detail because of lack of time.

(Q) What was the word you used when you described the slight lack of correspondence between the two spectra?

(A) Studier: Oh, isomerization . . . what I said was that in the old meteorite we didn't have an exact one-to-one correspondence with the synthetic. Maybe this methyl group, in a billion or so years, moved over, for example, so that we have it in a different spot. But it's still branched—we can tell by the discontinuity in the mass spectrum.

(Q) What is the chemical term for a branched hydrocarbon?

(A) Studier: Well, I think branch is a general term. Specific branches may be identified by numbering carbon atoms. For example, this is 2-methyl pentane

$$CH_3—C^N—CH_2\ CH_2—CH_3$$
$$|$$
$$CN_3$$

(Q) The range of disciplines represented here is very great. It's quite clear, to get the most out of what is being presented, you may have to ask questions which may seem to be fundamental to some people but are completely obscure to others. Are there other questions?

(A) Studier: I'd like to amplify my answer to the question about the time-of-flight. The advantage of the time-of-flight method is that the entire mass spectrum is displayed 10,000 times a second on the oscilloscope screen. So when the least little bit of material of any mass appears you can see it. This is a characteristic difference from the conventional mass spectrographs where you scan one mass unit at a time more or less slowly over the entire mass range. That is what makes the time-of-flight a good tool for the chemist. I'm not a mass spectroscopist; I'm a chemist, but I'm using it as a chemical tool. It's a very, very powerful one.

(Q) Although I'm quite familiar with the general principles of the time-of-flight spectrometer, I don't recall that you mentioned any details of the detector in the instrument.

(A) Studier: It's an electron multiplier. The ion strikes a stainless steel plate to produce electrons, which are then made to go down what is called a dynode strip. This particular machine has a unique type of multiplier. These are semiconductor strips. When an ion strikes, it produces electrons, which make cycloid paths because they are in crossed electric and magnetic fields. They keep striking the surface many times. There is a potential gradient in a direction such that every time the electrons strike they get one or two extra electrons to come out, so you get a gain of maybe $10^6$ going down the multiplier. In order to measure times of flight you have to pulse, so that at the time you pulse the ions out of the ion region you can set delay lines and other things in motion. You can take off, for example, the electron pulse due to a given mass as measured by the time from the start of the pulse. At this end you put a pulse on to draw this electron group off. It can also be made to scan so that the time at which you draw off the electrons due to the various masses will give the spectra that I've shown. For example, when you pulse you can start the sweep on your oscilloscope. When hydrogen comes in you get a current pulse at a later time and you get a voltage pulse. When water arrives there is a pulse at 18, for example; when nitrogen comes in, there is a pulse at 28, etc., all the way up to 2,000.

(Q) Is this detector commercially available?

(A) Studier: Yes. There are miniaturized versions of it that are used for very high gain multipliers. It's from the Bendix Corporation.

(Q) If an asteroid is indeed the source of meteorites as Mason indicated, would these asteroids be likely to have the primitive atmosphere that you are hypothesizing here in your synthesis?

(A) Studier: The organic compounds were probably formed before the asteroids were formed. Our idea is that they were formed in the earlier solar system before these asteroids condensed out. Initially it was very hot, and under those very hot conditions carbon monoxide and hydrogen are the stable species of those three elements. Somewhere along the line, when things were condensing out, conditions were favorable for the catalytic production of organic matter. Later on, they may have been reheated after, say, a large part of the hydrogen was lost. Incidentally, if I have a very high ratio of hydrogen to carbon monoxide, I do not form aromatics. No matter how hot I heat it, aromatics don't form. At very high ratios, approaching the cosmic ratio of hydrogen to carbon monoxide, we do form the straight chain at least out to 12 carbon atoms at a ratio of about 250 to 1. Later on, after the hydrogen had escaped, perhaps it was reheated to form the aromatic compounds that were found.

# Origin of Tektites

## JOHN A. O'KEEFE

*Abstract: Tektites are small green to black siliceous glass bodies which resemble obsidian but differ in chemical composition from terrestrial glasses. Although tektites appear in many different shapes, two facts are significant: their size is of the order of centimeters, and their surfaces are always pitted, grooved, or moulded. Prior to the Surveyor experiments, opinion on the subject of the origin of tektites was fairly evenly divided between the proponents of a lunar origin and the proponents of a terrestrial origin. The successful series of experiments conducted by Professor Turkevich on the Surveyors in September and November of 1967 and January 1968 have had a great impact upon the problems of lunar geology and thus upon the possible origin of tektites. The current thinking about their origin is discussed.*

INTRODUCTION

It's a great privilege to be here and to talk to so many people who have what I, as a licensed practitioner of astronomy, consider the background of scholarship—professors, Ph.D.'s, and all the rest of that. You have the greatest advantage which any university can have for the study of things which are out of this earth—that is, the fact that most of you have no big observatories. I think that few attitudes have done more to hold back the progress of studies of the moon

**Fig. 3-1. Examples of type B tektites showing characteristic pits and grooves. Many are teardrop in shape. From A. Lacroix, *Arch. Mus. Nation. Hist. Natur.*, 8, 1932.**

and the other planets than the feeling that every university should have a junior-grade Mount Palomar. It used to be said that everything we know about extra-terrestrial space came down the barrel of the telescope tube. Since the advent of research via satellites it has become clear that at least it is not true for them. In fact, for many years a good deal of information has been available to us in the form of meteorites, and this is a much more democratically available material than certain telescopes. Only the people experienced with very big telescopes can use such instruments, whereas almost anybody can obtain meteorites. At least some of the problems that can be done with meteorites are quite simple.

## TYPES AND DISTRIBUTION OF TEKTITES

There is a great abundance of tektite material and it is not terribly expensive. The present cost is about 5 cents per gram if you go to the right kinds of dealers.

Figure 3-1 is a picture of some tektites. They are not exactly typical, because hardly any picture of any group of tektites is typical, but it does suggest some of their most important features. Tektites typically have more or less droplike forms. This is a very, very significant property. Often they are more or less spherical, sometimes they are elliptical, sometimes teardrop-shaped. Many of these also have some characteristic grooves and pits which appear in the figure. The variety of the types of grooves and pits is so enormous that it's really very difficult to make an intelligent generalization. Tektites have a rather narrow size range, of the order of centimeters. It is very, very unusual to see a millimeter-size tektite; it's very unusual to see decameter-size tektites; and nobody has seen a meter-sized tektite. I shall call these type B tektites.

An example of the so-called Muong Nong tektite is shown in Figure 3-2. I shall call these type A. Externally these are rocky and uninteresting. If one were not a tektite student and were presented with a Muong Nong tektite, he would pay no attention to it at all. Typically, it has a layered appearance; it looks chunky, blocky. It never has the beautiful drop shape or spheroidal appearance of the type B. But it is the most interesting group of tektites if you slice them up and look at the interiors, because all kinds of things have been found inside these Muong Nong tektites. They are perhaps the most primitive, which is why I prefer to call them type A.

Type C is represented by the australites shown in Figure 3-3. These are three views of two different australites, but they are so nearly alike that they might as well be three different views of the same australite. Note the typical bowl-shaped form, the characteristic flanges on the outside, and the glassy

**Fig. 3-2.** Muong Nong, or type A tektites. Externally these are in sharp contrast to type B, but the internal structure is very revealing.

**Fig. 3-3.** Comparison of three Australian tektites (above) and three ablated models (below). Courtesy D. R. Chapman.

hemisphere. These features are almost certainly the result of descent through the atmosphere. Australites are plentiful, but specimens as perfect as these are very rare even in Australia where these forms are found. However, most of the imperfect Australian tektites are variants of this with a flange broken off or imperfectly developed. We understand this form rather well.

The distribution of "finds" of tektites is shown in Figure 3-4. In the United States, tektites known as bediasites, whose ages are around 33 million years, are found in Texas and, more recently, in Georgia. (There is an enthusiastic insurance salesman in Atlanta who found about half the national collection of Georgia tektites.) One was found in Martha's Vineyard. Apparently all tektites are about the same age, either very late Eocene or early Oligocene. There was a state-line fault between Texas and Louisiana which would be eliminated if one assumes that the Texas depositions were in fact very early Oligocene. [A state-line fault sometimes occurs along the boundary between two states as a result of a disagreement between their respective geological surveys in assigning ages.] These are about 33 million years old, anyway, whether Oligocene or Eocene. At about the same age is found the Libyan Desert glass, which may or may not be related to the tektite problem. For reasons to be discussed later I think they probably are related—but I emphasize this view is very far out, and quite controversial. The European tektites are about 15 million years old, which places them at Helvetian, which is late Miocene. The Ivory Coast tektites are about one million years old. The Australasian is a main group of tektites. Most of the tektites that have ever been seen come from this great field which extends from southern China all the way to Tasmania. Recently, some tiny tektites were found on the ocean's bottom, which extend this field all the way over across the Indian Ocean to Madagascar. The tiny microtektites are probably a part of the tektite problem, but again there are differences of opinion. The composition appears to be different also, so that in the case of the microtektites the geographic limits of the system, as well as the physical limits, are being extended. Naturally, this regional distribution is a great puzzle. If tektites are somehow from the outside of the earth, how did it happen that they are always found only in certain regions? This is one of the great problems. I will attempt to give an answer, but I should warn you that the answers have a certain ring of doubt or insincerity to those who are not devotees of the idea of an external origin.

PROPERTIES OF TEKTITES

The tektites can be considered in more detail, starting with type C, whose external form we understand the best; then type B, which we are beginning to

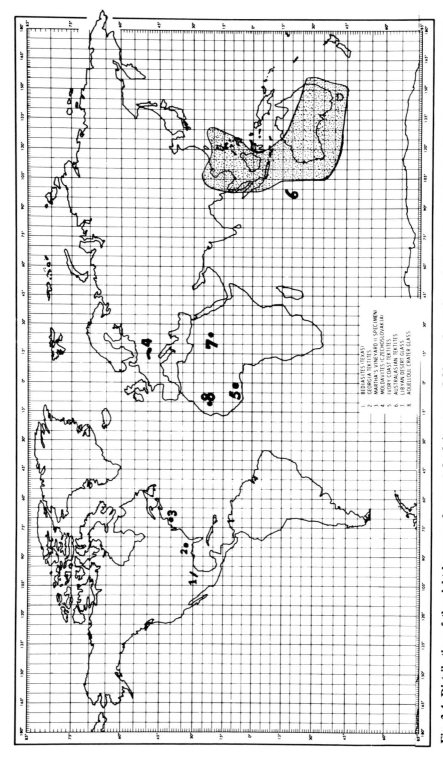

**Fig. 3-4. Distribution of the eight known and suspected tektite-strewn fields on the surface of the earth. 1. Bediasites (Texas). 2. Georgia tektites. 3. Martha's Vineyard (1 specimen). 4. Moldavites (Czechoslovakia). 5. Ivory Coast tektites. 6. Australasian tektites. 7. Libyan Desert glass. 8. Aouelloul crater glass.**

understand, although there is violent controversy about the meaning of the form; and finally, type A, which is the most primitive and which gives us a clue about the origin of the tektites.

D. R. Chapman, an experimenter at the Ames Laboratory of National Aeronautics and Space Administration, ablated samples of glass in the gas stream of a wind tunnel shown in Figure 3-5 and obtained forms almost iden-

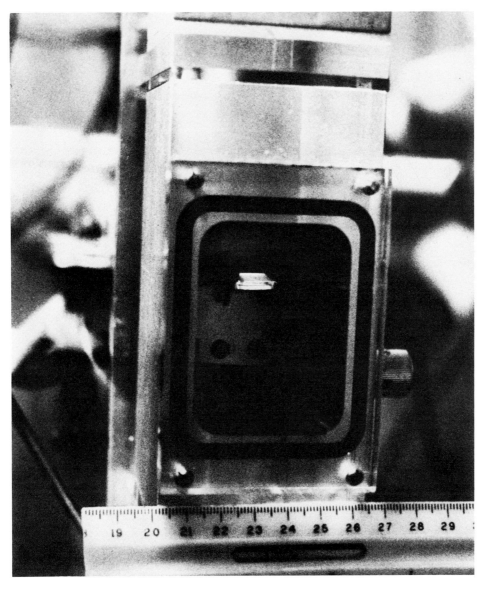

**Fig. 3-5. Chapman's chamber for producing synthetic tektite shapes with real tektite material.**

tical with the Australian tektites. These sections of these synthetic models of australites show the same internal structure of accordion-like foldings of layers of the more resistant glass. It is almost impossible to believe that the australites have not descended through the atmosphere at hypersonic speeds. The speed at which they descended was surely in the bracket for a body coming from the moon. Its velocity was between a minimum of 6.5 and a maximum of 11.5 km/sec. At velocities greater than 11.5 km/sec they get ablated away completely, and at smaller than 6.5 km/sec, even for the most favorable orbits which are the grazing orbits, the amount of ablation is far less than observed in the australites.

The most significant specimens of type B tektites were discovered recently by Nininger of the American Meteorite Laboratory. Searching through about fifty thousand of these rather cheap Australasian tektites, he found a few specimens that had the characteristic appearance shown in Figure 3-6. It appears as if it had been broken open while molten. This seems to be the most decisive piece of available evidence that the sculpturing was produced in the atmosphere while the tektite was still hot. The hypothesis is that the object came down through the atmosphere as a glob of molten material, that the external sculpturing was produced on it, that it broke open while it was still viscous, thus exposing part of the interior, and then it arrived at the surface of the earth, cold. It is very difficult to interpret this in any other way. It is important because it implies that the sculpturing of tektites is aerodynamic and not a

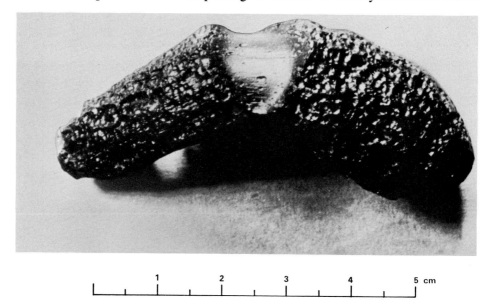

Fig. 3-6. Type B tektite. Bent while plastic. Courtesy H. H. Nininger.

result of ground-acid erosion. This interpretation has been a matter of violent controversy, and I warn you that the majority of students of tektites think that it is, in fact, ground-acid erosion. However, my point of view is shared by no less an authority than H. H. Nininger. The evidence strongly suggests that the sculpturing is almost certainly aerodynamic, if it is not the result of ground-acid erosion. If this is true, the tektites must therefore be the result of some mass of material which was at one point molten in the atmosphere.

Several of the facts can be accounted for if one assumes that a whole family of tektites is produced by the arrival of a single, large body which melts and drips in our own atmosphere to form the type B tektites. A few of the bodies, namely australites, which are very rare compared to the others, follow the main body, separate and fall on Australia. These bodies were once molten, then re-solidified, came down through the atmosphere as cold bodies, and formed the very peculiar, curious, type C forms. The australites fell very far away from the others, and they are the only kind of tektite found in Australia. This process could, of course, act as a natural filter. The only way they could have reached Australia is by having trajectories different from the others. It is not surprising perhaps that the difference in trajectory is reflected in difference in shape.

The type B tektites are enormously more numerous than the others, and they are probably much closer to the problem of the origin of tektites. Whether the sculpturing is aerodynamic in origin or due to ground-acid erosion has been a moot question for the past seventy years. The great man Suess, who first proposed the word tektite, was firmly convinced that the structure was the result of aerodynamic attacks. His strongest argument was based upon grooves or lines due to small inhomogeneities in the material which can be traced from one location to another in the tektite. But they are not the main feature which gives rise to these attacks. The main pattern of grooving and pitting pays only very slight attention to these chemical differences and is interested in something else entirely.

I once had a Japanese visitor in the office and I had five pictures of obsidian which had been corroded on the ground and five pictures of tektites. I laid them side by side and said to him, "Five of these were made by somebody who had no respect for the principles of Chinese calligraphy, somebody who made Chinese characters the way a Westerner makes them with a clumsy, awkward hand. The other five were sculptured by some force which sculptured things the way a Chinese does when he's making his Chinese characters—like a man. Please pick them out." He separated the tektites and obsidians without any trouble.

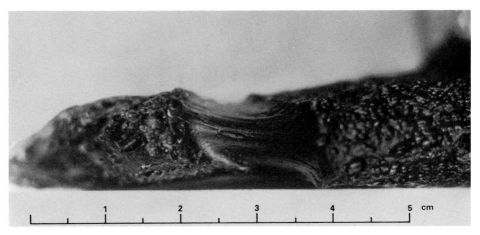

**Fig. 3-7. Type B tektite. Bent while plastic. Courtesy H. H. Nininger.**

The patterns suggest an aerodynamic force at work—something which is following a differential equation. It gives a firm stroke to the whole structure. In contrast the ground-acid erosion follows the wavering of the layering as if it is worried about irrelevant irregularities such as a microfissure or where a root ran; it gives an irregular, unhappy pattern to the whole thing.

In particular, one can state two empirical rules: The sculpturing on tektites tends to diverge from the center; and, if the tektite is an oblate spheroid with an identifiable equator, the grooves tend to go perpendicular to that equator. In addition to these rules there are other ways in which the sculpturing of tektites is sometimes sensitive to the form, the over-all shape, but insensitive to the internal layer. I suggest that this kind of sculpturing may not be due to ground-acid erosion. Here, again, is a problem which is quite accessible to people with modest means in a small department of a university. There is nothing to prevent one from buying 10,000 or 50,000 tektites and starting to look at them with no tools at all but a magnifying glass. This is what Nininger did, and he came out with rather interesting, surprising results.

The suggestion was made that the hole in the crust of the tektite in Figure 3-6 was due to the spalling off of a fragment right at the elbow. Figure 3-7 is another tektite on which it can be seen very plainly that spalling will not account for the situation because evidently this tektite twisted so that the axis of the left part is different from the axis of the right part—it twisted after the surface layer broke off, and that is impossible unless the thing was fluid. It is something like a Baby Ruth candy bar, with a sort of plastic interior and a more or less rigid outside.

At Marshall Space Flight Center a synthetic tektite made of real tektite

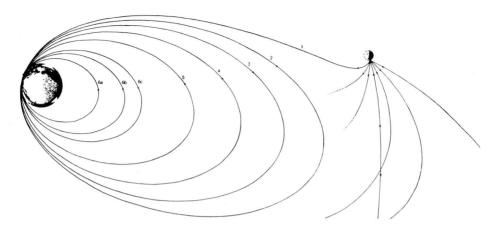

**Fig. 3-8. Illustration of the collision of a meteor with the moon. Only a small portion of the spallated material reaches the earth. By permission of *Scientific American*.**

glass was subject to aerodynamic attack in a kind of clumsy apparatus that the old-timers used to use when the study of ablation began. It produced ridges which resemble tektite sculpturing. (Marshall Space Flight Center is very much interested in flame dividers that split the exhaust and make it squeeze out on one side or the other.) Obviously right at the crest of the divider a material is required which is very resistant to attack. The important point is that, unlike the very refined people at Ames, Marshall used dense gases. The australites are produced, obviously, clearly, from bodies which entered the atmosphere from the outside as more or less rigid spheres, first encountering gases of very low density as they entered the atmosphere, thereby explaining the australite form. But to produce forms like the sculpturing, or forms like the little holes seen on the type B tektites, dense gases are required. A small body entering from the outside has no right to see that kind of ablation, because that kind of ablation is produced by turbulent flow. But the atmosphere doesn't give turbulent flow until a fairly high density is reached, and a small body entering from the outside will have its velocity arrested before it reaches a density at which turbulent flow is possible. In short, a small body entering from the outside will never see turbulent flow. So the presence of turbulent flow means that the body somehow got down to levels that it wasn't entitled to reach by itself. This suggests that it came in with a large parent body and plowed down to a level where turbulent flow is found. For example, really large meteorites show evidence of turbulent flow which produces the familiar regnaglypts. Although this is a fairly refined point, it begins to suggest that the tektites arrived as single bodies which were split up in the atmosphere.

ORIGIN OF TEKTITES

A most attractive hypothesis assumes that a meteorite strikes the moon as shown in Figure 3-8. This is a highly controversial picture and, for reasons which I shall discuss, it may not be the right one. After the meteorite collision, many fragments are produced, of which a very small proportion go into orbit around the earth. These orbits gradually decrease and they end in a grazing orbit in the upper atmosphere. At that point, the type B tektites managed to

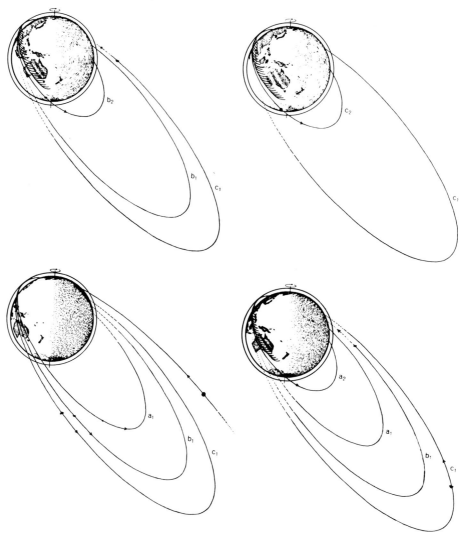

**Fig. 3-9. Successive orbits during which the earth turns can explain the longitudinal distribution of tektite-strewn fields. By permission of *Scientific American*.**

be produced. The problem is to explain why tektites are distributed over a wide range of longitude. Consider Figure 3-9 in the order of the letters. The body shown by the dot arrives initially along a trajectory as illustrated in the lower left of the figure. It swings around and breaks them into three bodies, $a_1$, $b_1$, and $c_1$. Body $a_1$ is more strongly retarded than body $b_1$ and body $b_1$ is more strongly retarded than body $c_1$. These are all fragments of this initial fragment from the moon. The right half of the figure shows body $a_1$ on its next pass around the earth as $a_2$, then it comes apart and strews Australia and the Far East with a strip of material as shown at lower right. In the upper left quadrant, body $b_1$ does the same thing. But body $b_1$ had a longer orbit than $a_1$, and therefore it arrived later at some distance away, and thus makes a strip which is further to the west, but parallel to the strip of body $a_1$. In the upper right body $c_1$ is now doing the same thing, but body $c_1$ had a bigger orbit than $b_1$, and therefore arrives later. The earth is turning in the meantime, and the strip has displaced even further. This can explain the fact that the Far Eastern strewn field is so widely spread out in longitude. This is in contrast to the explanation that the object simply came down with one big bang on the earth and spread fragments all over the Far East.

An example of an actual earth satellite which did in fact behave in this way is shown in Figure 3-10. This is what happened to the sustainer of John Glenn's satellite. When Glenn was put into orbit around the earth the vehicle

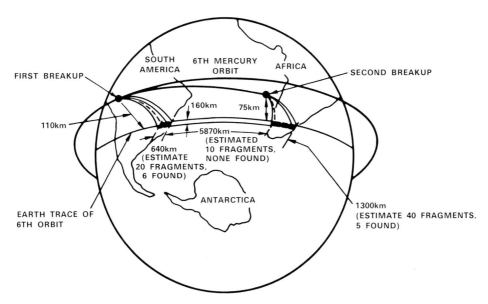

**Fig. 3-10. Estimated breakup and known recovery of fragments from John Glenn's sustainer stage.**

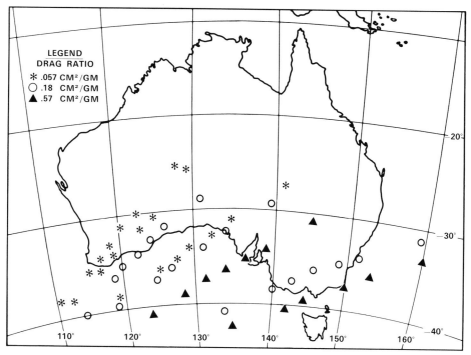

**Fig. 3-11. Distribution of computer-calculated fall-points for a fictitious strewn field superimposed upon a map of Australia.**

in which he was riding came down nicely, but the rocket that put him up there remained in orbit, of course. Later it came down, too, and for weeks afterwards natives were coming out of the jungle with pieces of stainless steel on their shoulders. In Brazil about twenty fragments fell, of which six were found. There were bits of the sustainer which were stamped, so that we were perfectly sure (1) that it was a U.S. satellite, (2) that it was an Atlas rocket, (3) that it was in fact the one that put Glenn into orbit, and (4) which part of the rocket it was. Another group of pieces came down in South Africa. Thus, with an orbiting satellite we can actually distribute pieces all over the earth. Of course, in successive orbits there would be a distribution in longitude. From the actual distribution it is inferred that the rocket broke up into two stages, the first at an altitude of 110 km, the second at 75 km.

The basic question is whether the localized distribution of tektites can or cannot be explained as the result of arrival from the outer atmosphere. To further investigate this question, Miss Shute and I made a calculation to determine whether, with reasonable conditions, we could cause the arriving bodies to be spread out. We found in fact that we could. We superimposed the map

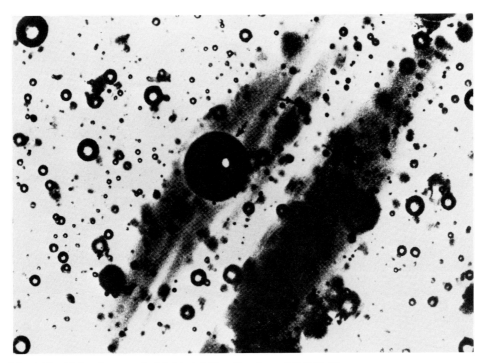

**Fig. 3-12. Ni-Fe spheres embedded in tektites. Courtesy of E.C.T. Chao.**

of Australia on it for comparison. It is clear from Figure 3-11 that one can get a distribution which is broad in latitude as well as in longitude by this kind of process. One can obtain numerical results with internal consistency.

We are fairly sure that tektites do in fact come as the result of an impact and are not volcanic. In Figure 3-12 the arrow points to a nickel-iron sphere embedded not in the Muong Nong tektites but in a type B tektite. A smaller nickel-iron sphere is indicated by the other arrow. This was discovered by E. C. T. Chao at the U.S. Geological Survey, who very kindly furnished the picture. The evidence is quite decisive. It is not only true that nickel-iron is present, but also present is the whole suite of trace elements found in the iron meteorites—the schreibersite and troilite (iron phosphide and sulfide). These things are without reasonable doubt actual particles of nickel-iron embedded in the tektite, which means that it was produced by impact on some surface somewhere.

One of these type A Muong Nong tektites was examined by Dr. Louis Walter in our office. Using a Chesley camera, he compared the x-ray picture with the x-ray picture of coesite, which is a natural polymorph of silica produced normally by impact which compresses the quartz, thereby trans-

**Fig. 3-13. Chesley camera x-ray photograph of coesite compared with a tektite sample. Courtesy of L. Walter.**

forming it into coesite. Figure 3-13 shows the coesite sample on the right and the tektite on the left. This was done in one of the tiny inclusions of $SiO_2$ which are spread through the Muong Nong tektites. Thus we have definite evidence of impact from the nickel-iron spheres and from the coesite.

In following the evolution backward from the type C tektites to the type B, and thence to fragments of the incoming type A tektites, we are now looking at data which result from the type A tektites and which tell us how they were initially formed. They were initially torn loose from something by impact. Did this impact occur on the earth or on the moon? Although the latest measurements from the moon did not show tektite glass, I still think that these are probably from the moon. You must listen to the argument very carefully and see if you really believe it. The strongest reason for believing that tektites came from impact either on the earth or on the moon, is the lack of evidence of a long sojourn in space. There is no evidence that the impact could have been on Mars, the asteroids, or some other planet because we never find in tektites any traces of the kinds of isotopes which are produced by the passage of a body through space. The primary cosmic rays produced the so-called cosmogenic isotopes. We don't find them. There remains only the question—was it impact on the earth or was it impact on the moon?

The most important characteristic of tektites is they are homogeneous glass. A slice through a Muong Nong tektite reveals the structure shown in

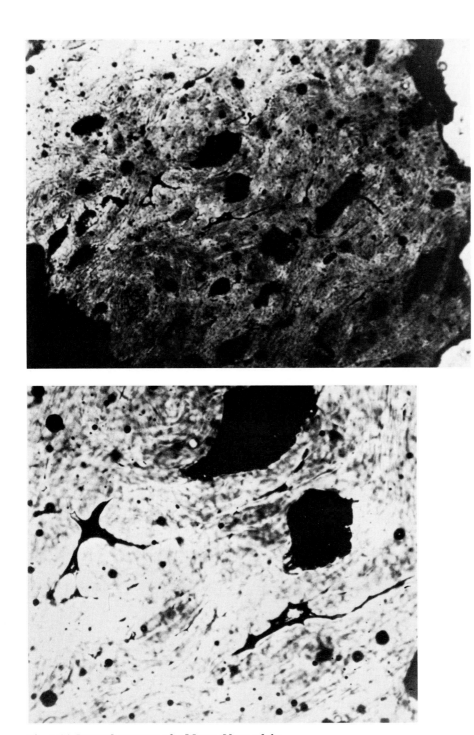

**Fig. 3-14. Internal structure of a Muong Nong tektite.**

Figure 3-14. There are discrete chunks of pure $SiO_2$ as well as voids. The tektite has been put together from tiny bits or shards of glass which have been pressed together while hot. The tiny bits are ignimbrite which surround the pure $SiO_2$. The coesite was found in these chunks. It has a rather inhomogeneous structure. This region has been examined by microprobe technique in several different elements, to illustrate the fact that although inhomogeneity in structure is obvious, there is chemical homogeneity. The material has been homogenized and was homogeneous already when this material was originally struck. It is evident that the material has not been physically melted; the voids between the separate fragments have been destroyed.

Figure 3-15 shows the extent of the homogeneity of $SiO_2$. Chemically it has a quite homogeneous structure. The voids are the two voids that appeared in the previous figure. A microprobe photograph of aluminum in the form of $Al_2O_3$ is shown in Figure 3-16. The compound that was used to polish the sample filled some of the holes in the $SiO_2$ nodules; there is quite a bit of porosity. In spite of its apparent inhomogeneity when looked at physically, it is in fact quite homogeneous chemically, which is not at all what is expected. If it was derived chemically from a terrestrial material at all, it must have been derived from a material which was about 60 per cent quartz sand, yet those quartz grains have vanished so that they cannot be found except for these nodules. It is evident that the silica nodules are in fact pure $SiO_2$ with voids. Microprobe photography of potassium and calcium are shown in Figure 3-17, again showing the very remarkable homogeneity of these materials.

The evidence poses a real difficulty for supposing that the tektite is produced from a terrestrial sandstone. The difficulty is that one must produce chemical homogeneity without destroying the physical inhomogeneity. We obtained some terrestrial material, the composition of which is very much like a tektite; it is a rather glassy material from Texas, apparently melted as a result of a lignite fire which underlay some sandy material. In this naturally fused material it can be seen that there is not the degree of homogeneity that is found in the tektites.

A silica nodule in glass photographed by phase contrast technique which show differences in refractive index is illustrated in Figure 3-18. Here is a mass of pure $SiO_2$ in tektite glass which has sharp boundaries; the material has not started to diffuse into the surrounding material. That is quite characteristic wherever the $SiO_2$ masses occur in the tektite glass—their boundaries are sharp. It makes it very difficult, at least for me, to believe that the tektites are in fact the result of the melting of a sandy material because it takes time for the sand to be melted and to be diffused into the surrounding material.

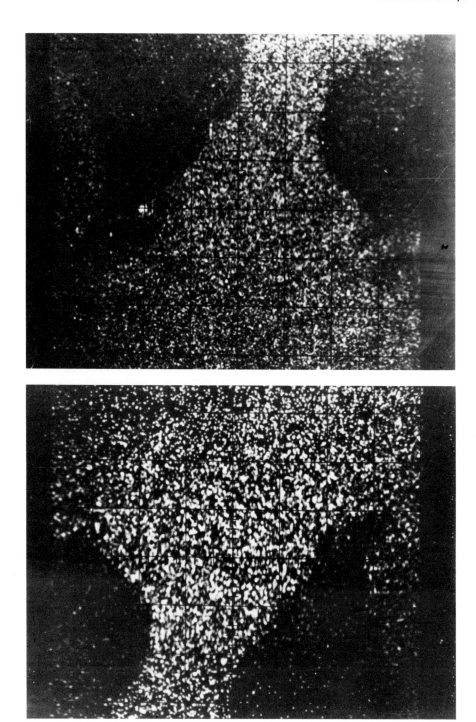

Fig. 3-15. Distribution of Fe (above) and K (below) in Muong Nong tektite (microprobe analysis).

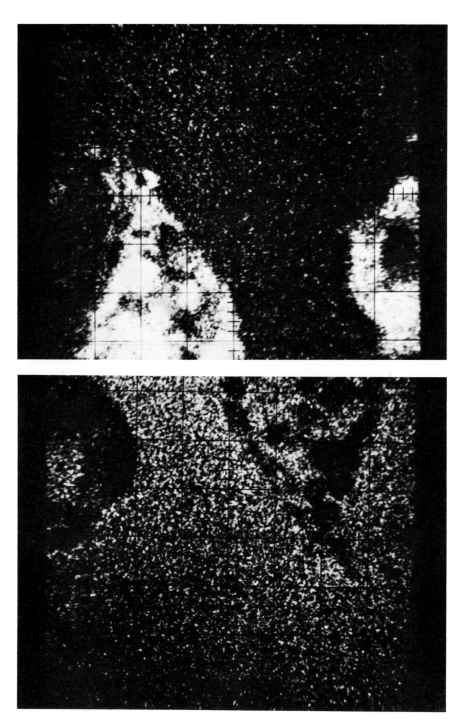

**Fig. 3-16. Microprobe scans for Al (above: grinding powder makes voids bright) and Si (below).**

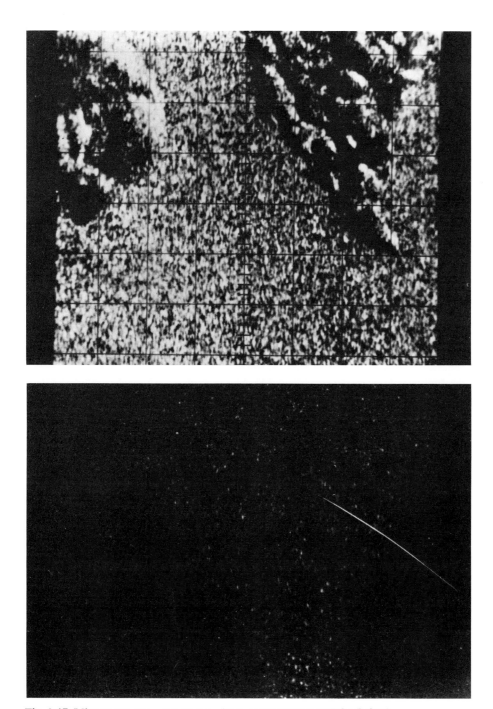

**Fig. 3-17. Microprobe scans for electron backscatter (above) and Ca (below).**

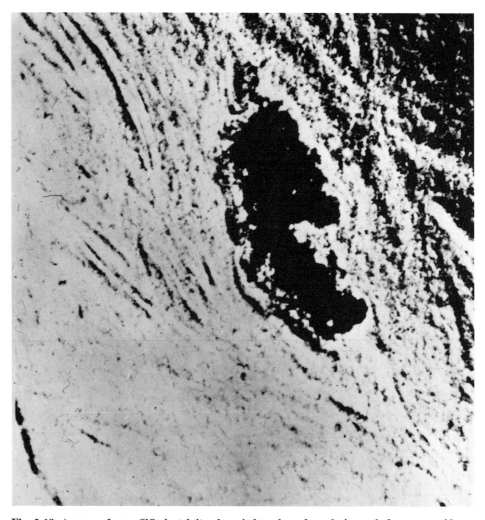

Fig. 3-18. A mass of pure $SiO_2$ in tektite glass; it has sharp boundaries and shows no evidence of diffusion into the surrounding material. The original measure of the picture represents .55 mm.

The ceramicists know what happens when a sandy material or a clay material is heated. Two problems appear. (1) Many bubbles and voids appear in the material—a tremendous number of them. The water comes out and forms these bubbles. The very bubbly material is not what is seen in the tektites. In fact, one sees in them fairly homogeneous glass. Take this problem around to ceramicists, and they will explain that the process of producing a fairly homogeneous material from a typical sand is *not* a sudden impact or any other rapid process. Elaborate precautions must be taken to get rid of the water; otherwise one won't get a glass out of any dirty old material one might happen to pick up. (2) The second problem they will point out is even more difficult than this process of leaching (i.e., to get rid of the water). It is the process of homogenization that takes place in the industrial plant. In the huge glass factories glass is produced in tanks called day tanks. The reason for the name is simple because it takes about a day to make glass in those tanks. If the tektites were produced by sudden impact on the earth, glass could not be produced in this way. Thus one is forced to conclude that the material was already glass before the impact took place. Since glasses like this don't exist on the earth, it probably is an extra-terrestrial material and very likely it is lunar. I warn you solemnly that there are other people who think differently.

Let us look now at the composition of the tektites and compare it with the composition of the moon. Typical tektite compositions are shown in Table 10 for 125 samples of seven groups of tektites; all show highly siliceous material —79 per cent $SiO_2$. The moldavites, for example, would pass for granite, except that if it had been a granite one would have expected that the $Fe_2O_3$ would have been more abundant and the FeO less. One would have expected the MgO to be less than half of 1 per cent. The calcium would ordinarily be around 1 per cent, although there are calciumless granites; sodium should be up to 5 per cent; potassium, titanium and manganese are about right. Water would be from 0.5 to 1 per cent in a granite, whereas in tektites the water content is always extremely low. Note that tektites are low in soda and low in potash. Dr. Mason pointed out that the eucrites are also low in soda, potash and water compared with terrestrial material. One of the most difficult problems in the study of tektite compositions is the analysis of volatiles. In 1959, I proposed that it would be very interesting to analyze tektites for chlorine. We made some contracts for analysis of tektites. Eight years (and I don't know how many million dollars) later, we have analyses for everything except chlorine and the other halogens. Apparently the basic reason it doesn't get done is because analysis for volatile materials in tektites gives negative results, so they are not reported. It is quite important to establish an upper limit. One basic chemical difference between tektites and terrestrial materials is in the volatiles. A published negative result with an upper limit would be very valuable.

COMPARISON WITH LUNAR MATERIALS

One of the important questions in the work with tektites is, how does their analysis compare with that of lunar material? In Table 11 the results from Surveyor V and Surveyor VI are shown in atomic per cent. One deep, fundamental result is clear. There is essentially no carbon. The oxygen combined with the silicon is about 50 per cent $SiO_2$, whereas the tektites have around 70 per cent. The magnesium was about 3 atomic per cent, which means about 7 per cent MgO; this, again, is much too high, compared with around 3 per cent for tektites. The aluminum was 6.5 per cent, which would work out to an atomic per cent of about 18. The agreement for $Al_2O_3$ is good, but the calcium and iron, however, are much too high. We find numbers which would be around 15 per cent CaO, and this means around 12 per cent FeO; they're

**Table 10. Mean Values of Compositions of 125 Samples of Different Groups of Tektites.**

|  | Moldavites | Billitonites & Javanites | Philippinites | Indochinites | Australites | N. America | Ivory Coast |
|---|---|---|---|---|---|---|---|
| N | 14 | 13 | 21 | 36 | 32 | 6 | 3 |
| $SiO_2$ | 79.01 | 71.29 | 70.87 | 73.09 | 73.06 | 78.59 | 71.05 |
| $Al_2O_3$ | 11.09 | 12.08 | 13.48 | 12.60 | 12.23 | 12.57 | 14.60 |
| $Fe_2O_3$ | 0.30 | 0.78 | 0.79 | 0.34 | 0.60 | 0.44 | 0.18 |
| FeO | 2.15 | 5.08 | 4.30 | 4.78 | 4.14 | 3.01 | 5.51 |
| MgO | 1.49 | 3.16 | 2.67 | 2.16 | 2.04 | 0.63* | 3.29 |
| CaO | 2.08 | 2.95 | 3.14 | 2.31 | 3.38 | 0.57* | 1.67 |
| $Na_2O$ | 0.52 | 1.57 | 1.41 | 1.45 | 1.27 | 1.34 | 1.71 |
| $K_2O$ | 3.04 | 2.20 | 2.31 | 2.40 | 2.20 | 2.13 | 1.53 |
| $TiO_2$ | 0.69 | 0.79 | 0.83 | 0.87 | 0.68 | 0.64 | 0.73 |
| MnO | 0.08 | 0.14 | 0.09 | 0.12 | 0.12 | 0.03 | 0.08 |
| Total | 100.45 | 100.07 | 99.89 | 100.12 | 99.72 | 99.95 | 100.35 |
| $Fe^\circ$ | 1.87 | 4.42 | 3.89 | 3.95 | 3.62 | 2.65 | 4.41 |

N = number of analyses averaged. * Two analyses from Barnes (1940) excluded for these oxides.

**Table 11. Chemical Composition of the Lunar Surface at Surveyor V and VI Sites (Preliminary Results).**

| Element | Percent of Atoms | | | |
|---|---|---|---|---|
|  | Surveyor V | | Surveyor VI | |
| Carbon .................................. | < 3 | | < 2 | |
| Oxygen .................................. | 58 | ± 5 | 57 | ± 5 |
| Sodium .................................. | < 2 | | < 2 | |
| Magnesium .............................. | 3 | ± 3 | 3 | ± 3 |
| Aluminum ............................... | 6.5 | ± 2 | 6.5 | ± 2 |
| Silicon .................................. | 18.5 | ± 3 | 22 | ± 4 |
| "Calcium" ............................... | 13 | ± 3 | 6 | ± 2 |
| "Iron" .................................. | | | 5 | ± 2 |

both much, much too high for tektites and clearly imply something of a general nature of a basalt. These analyses are for all of Surveyor V and Surveyor VI; three days ago, the data became available for Surveyor VII. Although I don't have a slide of the results, I have seen them and studied them carefully. In other words, from Surveyor VII again came something like a basalt, but not quite an iron-rich basalt. When my friends organized the place where Surveyor VII was sent, they chose a location where they hoped there would be tektites, if there were any anywhere. They didn't appear—there wasn't a real piece of evidence in the whole crop. It clearly shows that the idea that I had first, namely, that the *mare* surface has tektite-type glass is false; second, that the idea that the lunar uplands are all tektite-type material is also false. Probably the conclusion that can be drawn from all this is: if the tektites do come from the moon, they must come from some fairly restricted areas on the moon.

I would like to draw your attention to one final point related to a point emphasized by the master who began this presentation, the man who really does know something about the whole problem, Brian Mason. The compositions revealed by the Surveyors are really not very far away from the compositions of eucrite. Many investigators have really seriously considered the proposition that eucrites are of lunar origin. It is certainly clear that the moon is volcanic, so that the fact that tektites imply a volcanic origin, for example, is no bar to the hypothesis that they come from the moon. There is the further resemblance in that the sodium is rather low compared with any terrestrial basalt. It is low as it was in the eucrites as Brian Mason mentioned in his comparison of eucrites with chondritic meteorites. It does seem possible at least that the eucrites and the tektites come from somewhat the same environment. They have several things in common. The low soda, the low potash, the tendency to have $FeO$ instead of $Fe_2O_3$, and the low water content. What impresses me most is the absence of nickel. If the eucrites come from the moon it means that the surface of the moon is depleted in nickel, but if the surface of the moon is depleted in nickel, then probably the whole of the moon is depleted in nickel. How did it get so depleted in nickel? Ordinarily on the earth this means it was in contact with iron—if iron and rock are side by side, the iron gets all the nickel and carries it away. Where would the iron have gone? It would have had to go to the core of the moon, but the moon has no core. If we accept the old idea that the moon derived from the earth, the nickel that we should have expected in the moon is really to be found in the core of the earth. That suggestion is based upon the new results and is quite in contrast to the suggestions made by your distinguished first speaker.

SUMMARY AND CONCLUSIONS

Prior to the Surveyor experiments, opinion on the subject of the origin of tektites was fairly equally divided between the proponents of a lunar origin for tektites and the proponents of a terrestrial origin. The proponents of a terrestrial origin brought forward the fact that tektites are much closer in chemical nature to the earth's crustal rocks than to anything else in nature. The tektites are chemically nearer to terrestrial granites than they are to terrestrial basalts or terrestrial ultra-basic rocks. They are much nearer to terrestrial rocks than to established extra-terrestrial materials such as meteorites. The resemblance extends from the major elements right down through the trace elements with the exception of the volatile materials.

In the second place, there are two places on the earth where a large impact crater is found only a few hundred kilometers away from an important tektite deposit. In both cases, it has turned out that the age of the impact crater is close to the time when the tektites were deposited. One of these cases involves the Ries crater in Germany and the moldavites of Czechoslovakia. The other case involves the Bosumtwi crater in Ghana and the Ivory Coast tektites. In the latter case, resemblances of age include both ages since last melted and differentiation ages, that is, ages since the material received its present chemical composition. From these facts, some authors have suggested that tektites are a kind of impact glass. When a great meteorite strikes the earth, a small portion of the surrounding material is heated by the impact and converted to a glass. Such glasses are known around such undoubted impact craters as the Ries in Germany, Wabar in Arabia and Henbury in Australia. The undoubted impact glasses are usually very bubbly and heterogeneous. These authors suggest, however, that, under some circumstances, relatively solid and homogeneous glasses such as tektites may be produced.

On the other hand, the proponents of a lunar origin have pointed out that the aerodynamic shapes of some tektites, particularly those from Australia, imply descent through the atmosphere at hypersonic speeds. The minimum velocity of arrival appears to be on the order of 6.5 km/sec. This velocity is unexpectedly large for bodies which are being propelled from point to point on the earth. One would have expected velocities of only 4 or 5 km/sec. There is a distinct suggestion from the aerodynamic data that the velocities were at least big enough to put the bodies into orbit, which, of course, would be utterly inconsistent with a terrestrial origin.

The second chief argument of those who support an extra-terrestrial origin is the difficulty of producing instant glass. It is true that a sudden heating can transform many crystals into the amorphous form. These are therefore, in a certain sense, glass, but a tektite is more like what we are accustomed to calling a glass. It consists largely of a reasonably homogeneous mixture of a number of oxides whose ratios to each other do not correspond to those of any known or possible mineral. The production of a glass of this kind requires a homogenization of the mix. It is known that in ordinary glassmaking processes the homogenization takes longer than anything else.

A subsidiary difficulty of the same kind involves the water. If you start with an ordinary lump of dirt and heat it suddenly, you do not get a nice green glass. Instead, what you get is a foamy mass. The bubbles are caused by the water which has been expelled both from the pores of the soil and from the very molecules which make up the soil. In any normal terrestrial material heated to 1000° C. or so, the escaping water forms a volume considerably larger than the solid material. The whole thing is very familiar to ceramicists as a bisque, that is, a sort of spongy mass which is the first result of heating a clay.

Tektites are quite unlike the sort of spongy, inhomogeneous material which results from the sudden heating of soil or ordinary rock. They are moderately homogeneous, reasonably solid glasses. At any point, the variation of an oxide such as $SiO_2$ is not likely to exceed 5 per cent in the immediate neighborhood and is often less than 1 per cent. The bubbles constitute 5 per cent or less of the total volume.

Since a great deal of effort and thought have gone into the process of getting the bubbles out of glass and getting the glass homogeneous in factory melts, it is unlikely that anything important has been overlooked in this respect. Thus, it does not appear possible that a great impact could produce a very homogeneous glass in a very short time.

For these reasons, many people thought that the hypothesis of a terrestrial origin was implausible. They pointed to a complete lack of agreement between the chemistry of the glasses in the neighborhood of the Ries crater and the tektites from Czechoslovakia, and they tended to regard the Bosumtwi-Ivory Coast connection as a remarkable coincidence. It was the opinion of this group of people that tektites had been formed somewhere outside the earth. The most likely origin seemed to be the moon since there was clear evidence that tektites had not traveled very far through space. This evidence comes both from the lack of evidence of cosmic ray attack and from the localized distribution on the earth. Hence many workers thought it very likely that tektites came from the moon.

Such was the state of the argument in the summer of 1967. In September, the successful result of the experiments conducted by Professor Turkevich on Surveyor V demonstrated that the *mare* surface in Mare Tranquillitatis was a basalt. Two months later, a similar experiment in Sinus Medii showed an almost identical composition. In January 1968 a third experiment on the flanks of the great ray crater Tycho showed a composition which was again basaltic and differed from the previous two chiefly in the amount of iron. These experiments have had a great influence on everyone's thinking about the problems of lunar geology. In the first place, they showed plainly that the suggestion made by myself that the lunar seas (the *maria*) were covered with ash flows of tektite composition was wrong. In the second place, the failure to find silicic rock on the outside of Tycho undermined one of the principal arguments for silicic rock on the moon anywhere, namely, the argument that the differences in iron between *mare* and highland ground correspond to an isostatic relationship. The question logically can be raised now whether the idea that the tektites came from the moon should not be given up all together.

It is my opinion that in spite of these difficulties it remains likely that the tektites are of lunar origin. In the first place, one of the principal difficulties to the suggestion of a lunar origin for tektites has been the fact that tektites imply volcanism on the moon. Since the basalts are also volcanic rock, it is clear that to this extent the new discoveries remove an objection to the lunar origin of tektites. In the second place, there is a suggestion from the chemical composition of the lunar analysis that some of them, particularly those in the *maria,* resemble the kind of meteorites which are called basaltic achondrites. The basaltic achondrites are basalts which have fallen from the sky as meteorites. These basaltic achondrites differ from terrestrial basalts in some of the same ways that tektites differ from granites. The differences can be summed up in the statement that the basaltic achondrites tend to be poorer in some of the volatile materials than the terrestrial basalts. This includes particularly water, oxygen, and the alkali metals. It is puzzling, if these rocks really are from different places, that they seem to be saying the same thing about the source from which they came. It is easier to believe that the basaltic achondrites and the tektites have a common source.

Moreover, attempts to find the nearest terrestrial analogue of the basalts which have been detected on the moon often reveal that there are at least small quantities of granitic rock associated with these basalts. The processes which lead to the characteristic sequences of basalts often yield granitic rocks as the end members.

On the whole, therefore, it seems most likely to me that tektites will be found to correspond to *some* of the materials of the lunar crust. I think it most likely that they correspond to some intrusions of apparently viscous rock such as are seen in some ring formations. It is clear, however, that they do not constitute the bulk of the *mare* surfaces as I was saying a year ago.

DISCUSSION

(Q) I'm a little bit lost here. If a solid meteorite hit the moon, there must have been some melting in order to produce smaller fragments which became glassy, right?

(A) O'Keefe: The suggestion I'm making is that the moon's surface already contained glass.

(Q) These impact fragments are from the meteor?

(A) O'Keefe: These are impact fragments *not* from the meteor because the meteor would always have an entirely different constitution from these tektites.

(Q) Impact fragments from the moon's surface?

(A) O'Keefe: Yes, the impact fragments from the moon's surface itself, detached by the impact of the meteorites.

(Q) Then they come from the moon, into the earth's vicinity in a larger body which in turn breaks up?

(A) O'Keefe: Yes, that's the picture. I'm sorry that it is complicated, but apparently these complications are central to the problem. In other words, if you don't have this rather peculiar orbit, the fragment simply comes down and strikes the earth, makes a crater, and distributes a little glass around it. Now there is one case like that, the Aouelloul crater shown in Figure 3-5. This is a crater in which there has been an impact crater formed on the ground, and around the rim is some glass distributed which in many ways is like tektite glass. But in that case, the glass is peculiar looking; it looks like Muong Nong material which, as I said before, is utterly inconspicuous. It's the fact of the smoothness and the roundness of the tektites that brought them to the notice of civilization. They're rare; they're used as jewels in West Africa, Australia, Southeast Asia, and Czechoslovakia. It was at jewel exhibitions in 1892 that the whole problem came to the attention of the European geologists. Thus it is very likely that this less attractive raw material doesn't get the attention of the refined stuff.

(Q) So really we're looking at fragments of the moon's surface?

(A) O'Keefe: I think so. But there are many people who don't think so.

(Q) Is the time to come from the moon along the elliptic trajectory of proper duration to allow for the melting and yet not cooling sufficiently to crystallize?

(A) O'Keefe: I think that the melting occurred at the lunar surface, and not that they were melted as a result of the impact. The reason for thinking that the melting was not a result of the impact is that the coesites would convert to cristobalite if given time at a high temperature, so you must remove it from wherever it was without melting the glass. The coesite is formed at the impact —once you've formed it, you can never melt it again. If you ever melt it again, before you can cool it the coesite will vanish.

(Q) But the australites were melted.

(A) O'Keefe: Yes, but that is a second melting in the atmosphere, and they never contain coesite. Neither the type B or the type C tektites have ever been observed to have coesite. Chao, who found coesite in the first place, and I agreed that the one thing we really wanted to do with coesite was to find it in tektites. He found coesite in a meteor crater specimen which I loaned to him from NASA, with the hope of developing the criterion for discovering material produced by impact from the moon. He had lots of specimens of meteorites and looked through them carefully. He was the one who found the nickel-iron spheres, but he couldn't find coesite in the type B tektites. As a matter of fact, it was finally found in the type A tektites by Louis Walter. Although Chao never published it (you know, you don't publish the fact "I didn't find coesite in the type B tektites"), I was close to him and I know he couldn't find it. If Chao doesn't find something, there's a very good bet it isn't there.

(Q) There is a widespread belief that the lunar surface must be essentially devoid of potassium oxide and sodium oxide.

(A) O'Keefe: No, I think that tektites are very likely from at least some portion of the moon's surface; we now think an extensive portion. Tektites are short on potassium oxide; it's like obsidian except for about 5 per cent changes. You need to double the amount of $Na_2O$, double the amount of $K_2O$, make corresponding reductions elsewhere. To make it into a terrestrial rock, you must reduce the alkaline earths so that the CaO must come down from 2 per cent in the tektites to less than 1 per cent, and the MgO from 2 or 3 per cent in the tektites to less than 1 per cent in the terrestrial material in order to make a reasonable granite.

(Q) Have you postulated a mechanism whereby this material from the moon's surface can come out far enough to get past the neutral point, or minimum in the potential, so that the earth's gravitational field can bring it into the earth's surface.

(A) O'Keefe: Well, a typical meteorite travels at about 25 km/sec. The escape velocity of the moon is at about 2.5 km/sec. The energy goes as the square of the velocity, so the average meteorite has one hundred times as much energy per gram as required to remove material from the lunar surface. This means therefore that if a meteorite strikes the lunar surface you are very likely to have material escaping from it. The best studies of impact processes in the laboratory indicate that the moon is on the whole losing mass as a result of meteorite impact—that it loses about four times as much mass as the meteorites bring in. There is no problem at all about removing material from the moon.

(Q) Would it be something like a skipping stone effect?

(A) O'Keefe: No, the thing comes in like a hand grenade. It drops in, and then there's an explosion. Since energy goes as the square of the velocity and momentum goes with the velocity, the skipping stone phenomenon appears only at low velocity, where momentum effects count and the residue is strongly vectorlike. The momentum is predominant. As the velocity increases the momentum doesn't increase very rapidly, but the energy does, and so it becomes more and more like a hand grenade. It deposits a certain amount of energy and it blasts a hole; that is why lunar craters are so often round instead of elliptical, even though they're coming in at an angle obviously most of the time.

(Q) If you have a localized impact on the moon, it seems to me that the fragments which would be shot off with all kinds of velocities and in all different directions would not seem to be localized as they fell back on the surface of the earth. Why wouldn't it make a uniform deposit of these things, instead of having them localized at certain points on the earth?

(A) O'Keefe: That's absolutely correct. That's the point of these remarks. The point of arrival should not be localized. Most of the fragments that go out into the solar system will eventually be recaptured by the earth. They will come in and they will make craters presumably on the earth. They'll make craters like the Aouelloul which involves quite a lot of mass. A crater like that made by a glassy body coming in from space is much less conspicuous than a distribution of more or less droplike bodies. If you make a crater, you make a hole in the ground; you can distribute tons of material that way and nobody pays any attention to it. A lot of it vaporizes. If it comes in vertically, a lot more of it is just local in that region, so nobody finds it. If, on the other hand, you distribute the corresponding amount of material across the whole of Australia, in drops about the size that every native and everybody else wants to get, then the things will be very conspicuous and you'll find them very easily. The natives everywhere like to recover these things. We have in Texas people who were keeping them in cigar boxes and calling them black diamonds, before the WPA went down there looking for them. An old man there has a beautiful tektite and all the dealers have been trying to buy it from him. He won't sell it. Tektites are treasures when they've been through this rather unusual process that I describe. But if they come down straight out of the air they look like a piece of cinder and nobody wants them. Chapman went upcountry in Thailand looking for Muong Nong tektites. He talked to some of the Chinese tektite dealers because they sell them in Saigon for jewels. The man showed him a lot of type B tektites, but Chapman showed him a Muong Nong tektite and said, "I want something like this." The Chinese dealer looked at him: "You don't want to buy that!" They're hard to collect if they're rough.

(Q) What causes the breakup when it reaches the earth's surface?

(A) O'Keefe: It passes through the atmosphere, and at considerable depth in the atmosphere the forces on it are tremendous.

(Q) That causes the fragmentation?

(A) O'Keefe: Tears it apart, mechanically and thermally.

(Q) The surface of the moon is, of course, subject to cosmic ray bombardment. Would you expect to find cosmic ray effects in tektites?

(A) O'Keefe: You certainly would. If these things came from the top surface of the moon they would have shown cosmic ray effects. We now know that the top surface of the moon down to a depth of at least a foot or two is finely divided material. We also know that if you dig down to a certain depth of the moon you begin to excavate solid chunks, so we think that probably this cosmic ray-contaminated upper layer is finely divided and escapes that way.

(Q) A breakup like the Glenn sustainer doesn't happen just over land; doesn't it happen over the ocean also?

(A) O'Keefe: Surely, but nobody ever recovers anything. These are actual pieces being actually recovered; these are pieces which were brought in by the natives. If something falls into the sea, nobody's going to find it, probably. That's the sad thing about metorites.

(Q) But is there any real reason for postulating that there were two explicit breakups rather than one?

(A) O'Keefe: Oh, yes. A very detailed analysis was made to show why there were two complete breakups. I couldn't reproduce it for you. The end of the thing which contained the jet exhaust was heavier than the rest, and so it survived longer than other parts.

(Q) From your picture, Figure 3-10, it appears that the second breakup didn't actually give any recoverable fragments; is that right?

(A) O'Keefe: No; some fragments came from the second breakup.

(Q) You said that these tektites probably derived from some kind of restricted environment on the moon. Do you visualize a surface or a subsurface environment?

(A) O'Keefe: Surface environment, probably an outflow of some viscous material.

(Q) In other words, this meteorite was probably very particular in its impact area on the moon's surface, right?

(A) O'Keefe: Well, it brought us the tektites and the other meteor impacts bring us the eucrites. I think of the eucrites also as being a secondary material from the moon.

(Q) By restricted environment do you mean only a few square miles?

(A) O'Keefe: I don't think we're in that type of situation. We have now three samples, none of which resemble the tektites, but there is a great deal of moon surface we haven't sampled yet.

(Q) The objects which Chapman ablated—were they glass?

(A) O'Keefe: These are tektite glass; they actually took a tektite and ground it into a sphere.

(Q) What is the definition of a tektite?

(A) O'Keefe: Small, glassy bodies, free of tiny crystals which distinguish them from obsidian, having no relation to the ground on which they are found, usually with a highly acid composition, corresponding roughly to a granite or something similar, and nearly always essentially water-free.

(Q) What is the size of the australites?

(A) O'Keefe: These are about centimeter size.

(Q) What is their age?

(A) O'Keefe: About 700,000 years; this is determined by argon dating, as well as fission track dating.

(Q) Is it true that the tektite distribution shows a very great concentration in the southwest Pacific?

(A) O'Keefe: Right.

(Q) Which is a volcanic region? Does this suggest a source other than extra-terrestrial?

(A) O'Keefe: No, this is really pretty solid. Velocities of 6.5 km/sec are not found in connection with a volcano. Moreover, embedded in some tektites are nickel-iron spherules which make it quite sure that it has something to do with impact eventually. And besides that, there is coesite in some of them—as I'll show you in a minute or two—that also indicates clearly that impact had something to do with it. Volcanoes just don't give you a velocity specifically distributed over an area like that. There are, furthermore, all kinds of volcanic glasses. The chemical composition of tektites is not too unlike a volcanic glass, but they are characteristically short of soda and potash; as a result they do not fit the characteristic composition of volcanic glass which always has 5 per cent or so soda and potash.

(Q) I noticed on the map no indication that tektites were found above 45 north latitude, or approximately 45 south latitude; is this because there haven't been systematic searches for tektites or is there some real effect here?

(A) O'Keefe: Half the earth's surface is between 30 north and 30 south, though you wouldn't think it to look at a map. When you look up to 45 you've included a major part of the northern hemisphere and north of 45 there are the tektites of Czechoslovakia, which are, I think, about 50 north. The tektites of Tasmania are pretty close to 45 south. So I don't think too much significance can be attached to it. Remember, north of 45 in the United States will take you up into Canada. So searches are certainly not as enthusiastic up there. Besides that, the ice age has ruined the terrain.

(Q) Will the time to go through the atmosphere in the test-chamber equal the time in flight?

(A) O'Keefe: I think that they were not able to reproduce all the details; they were not able to model that way. It was necessary to make some change in the chamber to take account of that. It's not possible to reproduce accurately because you cannot get the velocities of Mach 25 in the tube. So they use a denser gas and a scale factor. The modeling was aerodynamically correct, as far as Chapman could make it. Since Chapman is one of the people who designed the re-entry cones, he probably knows what he's talking about.

# II. Lunar Science

# The Lunar Regolith

## EUGENE SHOEMAKER

*Abstract: The Surveyor Program has produced a great amount of pictorial
data of unprecedented resolution of the lunar surface, which provide
the basis for a quantitative description of the lunar regolith.
On the lunar maria the observed regolith varies in thickness from 1 to 3 m
to more than 10 m and is composed of fragments whose mean size is
less than 1 mm. This layer of debris is believed to have been formed by
repetitive impact, which also produced the majority of the observed
small craters.*

INTRODUCTION

It is a pleasure to be with you and to tell you some of the most recent results
obtained from the U.S. Lunar Flight Program. I will confine my remarks to
some of the new things that we have learned from the Surveyor project about
the debris layer of the lunar surface. The moon has a layer of debris similar
to that on the earth, but it covers a larger portion of the moon than the earth—
about 99 per cent of it as far as can be judged from our Lunar Orbiter pictures.
I shall briefly describe this debris layer and then present a model of how I

**Fig. 4-1. Photo mosaic map of the moon showing the most suitable possible landing sites for project Apollo.**

think it came to be formed. I call this debris layer a regolith; it is an old word for similar layers of debris that cover up the bedrock on the earth. Although the term has not been widely used, I think it is perfectly appropriate for the layer of debris that I'm about to show you.

The photo mosaic map of the moon shown in Figure 4-1 identifies the region along the equatorial belt of the moon that was selected for reasons of celestial mechanics as most suitable for possible landing sites for the men in Project Apollo. A major part of the Surveyor program was devoted to landing unmanned spacecraft on the *maria,* the smooth plains areas of the moon in this equatorial region. Such areas seem to offer the best landing sites for the first manned lunar landing. Four spacecraft successfully landed on *mare* surfaces within this belt; Surveyor I and Surveyor III landed in the Oceanus

Procellarum, which is the largest *mare* area on the moon. Surveyor V landed along the southern margin of Mare Tranquillitatis, and finally Surveyor VI violated Shoemaker's law (which says that only odd-numbered spacecraft can land) and landed on Sinus Medii, near the center of the moon.

The surveyor spacecraft is shown in Figure 4-2. It is sitting here not on the moon, but on the shore of "Mare Pacificus," not far from the Hughes Aircraft plant where the beast is manufactured. To give some idea of the scale, the footpad is just 1 ft across. The spacecraft is a three-legged monster with its legs and shock absorbers coming out from a triangular frame. Mounted on the frame is a mast that carries the planar array antenna, a narrow beam antenna over which the wide-band television signals were sent, and the solar panel with solar cells for converting sunlight into electrical energy to power the electronics compartments that are mounted on the back. I do not have time to describe the multitude of fascinating details; the most important feature for this discussion is the television camera. On the top of the camera is a little rotating turret, which has a mirror in it that can be tipped up and down. The turret can be swung around through a nearly 360-degree azimuth. Below

**Fig. 4-2. Surveyor VI shown on the shore of "Mare Pacificus" (original slide in color).**

**Fig. 4-3. A close-up photograph of the proposed landing site of Surveyor I taken by Lunar Orbiter I.**

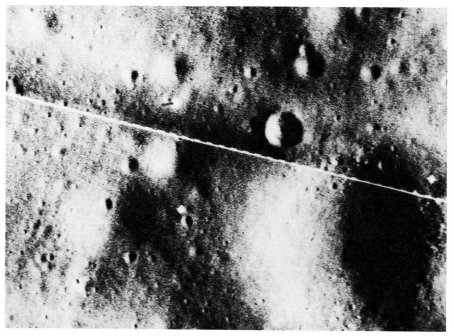

**Fig. 4-4. Surveyor I after landing on the moon is shown in this photograph taken by Lunar Orbiter III.**

the turret is a zoom lens with variable focal length ranging from 25 to 100 mm. At the base of the camera is the electronics part with a vidicon tube. The focal plane is at the target of the vidicon tube. We were able to take pictures with angular fields of view ranging from 6° in the narrow angle or long focal length mode, up to 25° with the wide angle or 25 mm focal length mode.

## SURVEYOR I

A closeup of the region around the first landing site, that of Surveyor I, is shown in Figure 4-3. This picture was taken by Lunar Orbiter I. It shows part of a discontinuous ring of low mountains that surround the region. This ring seems to be the exposed parts of the upper rim crest of an ancient crater which has been largely buried by *mare* material. On the *mare* surface itself a rather typical array of small craters is seen in the photograph. They are too small to be visible with a telescope and were first seen in the Lunar Orbiter pictures of this part of the moon. To give an idea of the scale, the crater marked with an arrow is a kilometer across. The other smaller craters have general dimensions of 200 to 300 m, and X marks the place where Surveyor I touched down. After Surveyor I landed, the site was overflown again by Lunar Orbiter III, and pictures were taken in a deliberate attempt to photograph the landed spacecraft. The successful result is shown in Figure 4-4. The spacecraft is the little white blob. It is not resolved fully in this Orbiter picture taken from a height of about 25 km with a 22-in. focal length lens.

Close examination of Figure 4-4 shows a little black stripe, which is the shadow cast by the planar array antenna and solar panel that were tipped up vertically in order to cast a maximum shadow in the early lunar morning when this picture was taken. Some of the features that can be seen from the camera on the Surveyor that also appear in this photograph include a rather prominent and sharply formed crater about 30 m in diameter which lies a little over 60 m from the landed Surveyor. One can see the near rim of this large crater and the far rim along the horizon from the Surveyor camera. The rim of another crater about 200 m across and a large number of fragments along the rim that are not resolved in the Orbiter picture could be seen along the horizon from the Surveyor I camera. Some of the coarsest fragments, as much as 5 m across, associated with this crater are resolved in the Orbiter picture. Two rather subdued craters are visible. One crater with a little raised rim is of particular interest; it is 9 m across and provides some critical evidence on the thickness of the regolith or debris layer at this landing site.

**Fig. 4-5. Part of the lunar landscape seen by Surveyor.**

Figure 4-5 shows the view as we saw it from the Surveyor I. The horizon is hump-shaped here because the Surveyor I camera was inclined about 16° from the vertical axis of the spacecraft and the normal to the lunar surface. We put this mosaic together by pasting the pictures onto a cylindrical projection, the axis of which is coincident with the axis of rotation of the turret of the camera. The horizon under these conditions, if it was perfectly level and smooth, as it very nearly is, would be a sine wave with a 16° amplitude. One of the distant peaks in the mountain ring about 12 km from the landed spacecraft can be identified on the horizon. The 30-m crater, which I mentioned earlier, can also be seen.

In the foreground are the kinds of features which were seen in good detail for the first time with Surveyor I. One could see similar features with lower resolution in pictures taken from Lunar IX, launched by the Soviets, which was the first successful unmanned lander on the moon, but these features were not at first well understood. Craters can be seen scattered everywhere, ranging from 2 cm up to tens of meters across. In addition, a close look reveals a large number of little bright objects scattered about the surface; we are

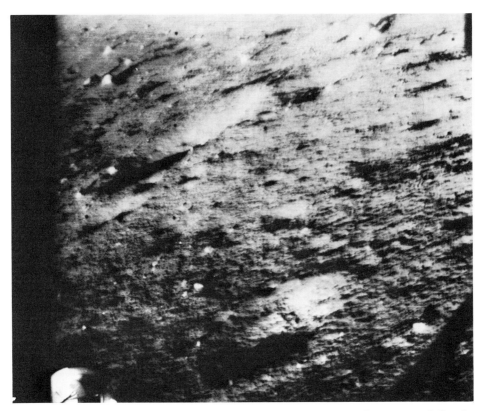

**Fig. 4-6. Wide-angle photograph which shows that for small craters the texture of the rim material is similar to the texture of the surface.**

fairly sure that these are rocky fragments. Some interesting details are the omni-directional antenna on the spacecraft, one of the footpads, the shadow of the spacecraft itself, and the shadow of the camera. This mosaic was prepared from pictures taken very late in the lunar afternoon.

Figure 4-6 shows some of the features in great detail. This picture is one of the wide-angle pictures used in the previous mosaic. One of the craters low in the picture is about a foot across and has a rather typical raised rim. The rims of these small craters and the crater walls have a texture similar to the texture of the surface between the craters. Bright angular chips and fragments are scattered about on the crater rims in about the same proportion as elsewhere. From the similarity in texture of the rim material to the other parts of the surface, I infer that craters have been excavated in material with essentially the same texture as that seen right at the surface. In other words, the fragmental material observed at the surface has an extension in depth at least as great as the depth of these craters.

**Fig. 4-7. The same evidence of similarity between rim and surface textures is found for larger craters of the order of 1 m.**

Similar relationships are found at larger craters. Figure 4-7 shows a crater about 10 m from the camera; it is $3\frac{1}{3}$ m across and 1 m deep. The picture was taken very late in the lunar afternoon, and it shows a low but well-defined raised rim around the crater with little rocky fragments scattered on it. For the most part, the texture of this rim is very much like the texture of the surface between the craters, so again I infer that this crater was excavated in fragmental material like that seen at the surface, and that the material extends at least to the depth of the crater, about 1 m. Craters of this size and smaller formed experimentally have rims this shape and form only if the debris in which the crater is excavated has very low cohesion. Figure 4-8 illustrates a bomb crater produced on the Marine target bombing range in

**Fig. 4-8. A bomb crater in the Mojave Desert. Many of its details are almost identical with moon craters.**

the Mojave Desert. It is just about the same size as the lunar crater shown in Figure 4-7, and it has very nearly the same shape, too. This crater again is formed in alluvium on the edge of a small playa. An upper limit of $10^4$ dynes/$cm^2$ can be placed on the cohesion of the layer of fragmental material on the lunar surface on the basis of the fact that the small craters have this kind of shape. The cohesion is comparable to that of very slightly damp sand. If the cohesion were higher, the crater rims would be conspicuously lumpy or blocky.

Some of the smaller craters seen late in the lunar afternoon are shown in Figure 4-9. Low sun angles are required to see these craters and to pick them out from the other features of the lunar surface. The craters shown here have diameters ranging from a few centimeters to about 15 cm. In some of the craters little bumps can be seen sticking up. A good analogy to these lunar craters is shown in Figure 4-10, which illustrates part of the secondary crater swarm around an impact crater at White Sands Proving Ground, New Mexico, a crater that was formed by the impact of a dummy ICBM warhead launched from Green River, Utah. The parent crater, several meters across, was

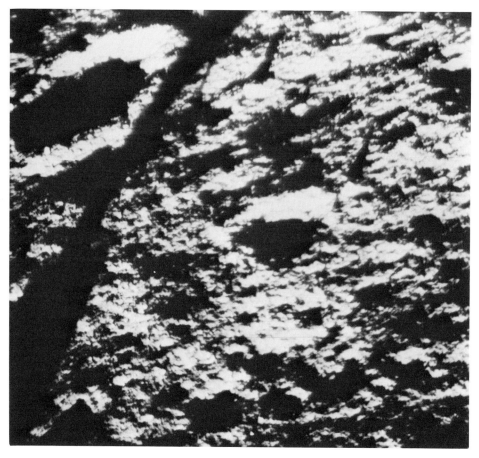

**Fig. 4-9. Some small craters on the moon seen late in the lunar afternoon.**

formed in material of relatively low cohesion, but some of this material was shock-compacted as it was ejected and formed harder little lumps which struck the surface. Some of the lumps are still resting in the craters they formed because these are very low velocity impact craters. They have been formed in the uppermost layer of the alluvium at this site, again in material of a rather low cohesion. I believe that many of the small craters we see around all the landed Surveyors are produced by impact of little bits and chunks and compressed wads of lunar debris that have been thrown out of other craters. Less than half of the craters should be of primary origin, that is, formed by interplanetary debris striking the surface, and more than half by impact of material from the moon itself.

Looking farther away at larger craters, Figure 4-11 shows a 9-m-diameter crater with a distinct swarm of fragments on its rim. This crater is 2 meters deep. I infer that the crater has been excavated down, partly into an underlying more cohesive, or at least coarser, substratum from which these blocks have been derived. All craters at the Surveyor I landing site larger than this have blocky rims—that is, if they have a raised rim. Figure 4-12 is a mosaic of high resolution pictures taken very nearly at lunar sunset, showing the 30-m-diameter crater seen in Figure 4-4. The rim is littered with blocky fragments

**Fig. 4-10. A secondary crater swarm around a primary impact crater at White Sands formed by a dummy ICBM warhead.**

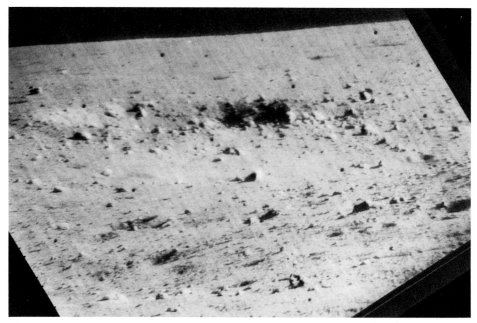

**Fig. 4-11. A closer view of the 8-m-diameter crater seen in Figure 4-4. Note the distinct swarm of fragments on its rim.**

**Fig. 4-12. The 30-m-diameter crater seen in a mosaic of high-resolution pictures taken near lunar sunset.**

which extend along the horizon an angular distance about equal to the angular diameter of the crater. A close-up of part of that rim taken at high sun appears in Figure 4-13. The larger pieces in this picture are about ½ meter across, and some of them are very angular. For example, by looking closely at one rhomboid-shaped fragment that sticks out, a little bit of the sky can be seen underneath one of the protruding edges. This crater is about 8 m deep, and it has been excavated well down into the underlying rocky substratum.

Some of the fragments close to the spacecraft may be part of the strewn field of ejected blocks associated with this crater. A fragment about a half a meter across that lies about 5 m from the camera (Figure 4-14) may be one of them. This fragment is rather angular, but its edges are distinctly rounded. It is spotted or freckled and surrounding fragments are spotted; in fact, there is a little strewn field, or patch of about 50 fragments, which have similar spots. I would infer that these fragments have been derived from a larger fragment which broke up on landing at this place. One of the most interesting things about the spots is that the light material sticks out as distinct little knobs or bumps. This tells us two things. It demonstrates, first, that some mechanical difference exists between the light material and the darker matrix, and, second, it strongly implies that some sort of erosion process has been going on, and that the light parts stick out because of differ-

Fig. 4-13. A close-up of part of the rim of the 30-m-diameter crater taken near high noon. Note the larger angular fragments which are about 1 m across.

**Fig. 4-14. A crater about 10 m deep excavated into the underlying rocky substratum.**

ential erosion. A number of fractures and fissures cut down into the surface of the largest spotted rock. The fissures appear to have been enlarged by erosion.

Features seen on other fragments also suggest differential erosion has taken place. A fragment about 15 cm across, highly rounded, and partly buried in the lunar surface, is shown in Figure 4-15. The feature of interest is the little ridge that runs along the top of the fragment. A feature like this is probably the result of differential erosion. It suggests there is a layer of resistant material within the block. Another rounded object, half a meter across, is shown in Figure 4-16. The most interesting feature of this rock is the host of little black spots scattered all over the surface. We were able to observe these spots through about 165° of travel of the sun during the lunar day and to show con-

**Fig. 4-15. This strongly rounded fragment is about 15 cm across and lies partly buried in the lunar surface; there is evidence of differential erosion.**

clusively that these little dark spots are shadows in deep cavities on the surface of the rock. These cavities are the broken or open ends of vesicles or gas bubbles in the rock and their presence indicates the material has been melted at some time in its history. Probably this rock is a piece of scoriaceous lava.

### SURVEYOR III

Surveyor III landed about 400 miles east of Surveyor I, in the eastern part of the Oceanus Procellarum, about 75 miles from the fairly large lunar crater Landsberg (Figure 4-17). Good high-resolution photographs were taken with Lunar Orbiter III before the Surveyor III landing, and just enough features were observed in the Surveyor pictures that could be correlated with features seen in the Orbiter pictures to locate the position of the landed spacecraft; it is shown by the white triangle which is the exact size of the spacecraft (Figure 4-18). Surveyor III landed in a subdued crater about 200 m in diameter and about 15 m deep. It has a rounded but gently raised rim and is a type of crater that is rather common on the *maria*. The spacecraft landed about halfway down from the rim to the center of the crater.

**Fig. 4-16. A large block close to the spacecraft. It has fractures partly filled with fine-grained material and many little black spots on the surface.**

There were some difficulties with the landing program of Surveyor III. The engines did not turn off at a few meters above the lunar surface as they should have from the signals of the radar altimeters. Instead, the spacecraft landed and hopped back up into space, then landed again with the engines still going, and hopped back up again, and finally the engines were turned off by ground command. We do not know how long the spacecraft could have gone on hopping around, but we got three landings for the price of one and learned quite a bit about the interaction of the footpads with the surface. We were surprised that the spacecraft had landed safely when we looked around to see the sort of boulder field in which it had been hopping.

Figure 4-19 is a Surveyor picture of one of the features we could identify in the Orbiter photographs. It is a crater about 13 m in diameter lined with

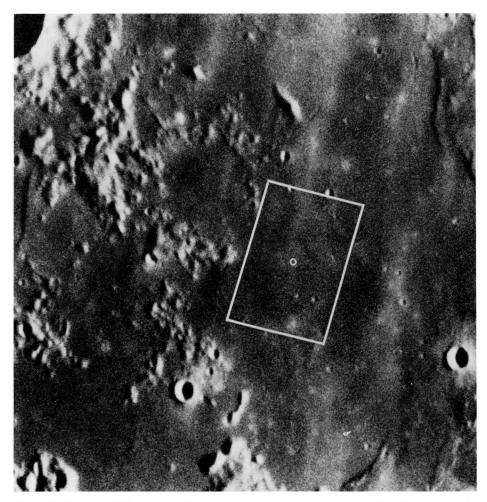

**Fig. 4-17. Surveyor III landed at this site in the eastern part of Oceanus Procellarum near the lunar equator.**

angular fresh-appearing blocks; the crater is surrounded by a swarm of blocks. Blocks ranging in size from about ½ m to about 4 m across could be recognized in the Orbiter photographs. Another crater about the same size was observed which is much more subdued but also has a block field.

We prepared a topographic map, Figure 4-20, of the main crater in which Surveyor III landed from the information obtained from features that could be correlated between the two sets of pictures—the Orbiter pictures, and the Surveyor pictures. About two hundred points could be so identified. The position of the spacecraft was found by triangulation. The distances of objects from that position to other points on the surface were obtained from the

**Fig. 4-18. The Surveyor III landing site as photographed by Lunar Orbiter III.**

Orbiter picture and the vertical angles were determined from the Surveyor camera. Thus, this map was constructed by methods very closely analogous to those used in ordinary field surveying. The contour lines are at 1-m intervals.

Among the most interesting features are the strewn fields of blocks associ-

**Fig. 4-19.** This 12-m-diameter crater lined with very angular fresh-appearing blocks was one of the many features used to correlate with the Orbiter photographs.

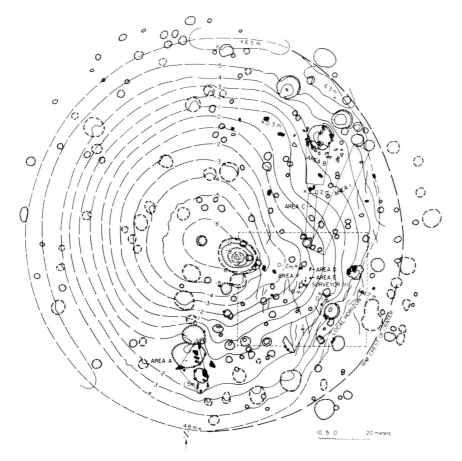

**Fig. 4-20.** Topographic map of the Surveyor III site.

**Fig. 4-21. One of the strewn fields of blocks on which measurements of angularity and rounding were made.**

ated with two craters shown in Figure 4-20, one crater about 16 m in diameter on the south wall of the main crater (area A), and the other 13-m crater near the northeast rim of the main crater (area B). One is a subdued crater and the other is a sharply formed crater. One of the more obvious differences between these two strewn fields of blocks is the angularity of the blocks as shown in Figure 4-21.

We attempted to measure the rounding of the blocks by the straightforward method of fitting circles to the corners of a block profile, taking the geometric mean of the radii of these circles, and then dividing it by the radius of the circle that just encompasses the block. This quotient we called the roundness factor. Figure 4-22 illustrates the frequency distribution of the roundness factor in the two strewn fields. Blocks around the sharply formed crater have very low roundness. Around the subdued crater the mean roundness of the blocks is significantly greater. These data provide further evidence that some process of abrasion or rounding is taking place that grinds down the surfaces of rocks on the lunar regolith.

Another feature of interest is the extent to which the blocks tend to be buried or embedded in the surface. In Figure 4-23 may be seen a rock about 15 cm across that is sitting right on the surface. Others are buried to varying degrees. Figure 4-24 shows a rock about half buried. Some unusual wedge-shaped fragments with just a corner or edge protruding (Figure 4-25) were also observed. The degree to which a block is buried can be measured from its profile against the more distant lunar scene. On each side of a block the angle is measured between the tangent to that profile, where it meets the surface, and a horizontal line. The sum of these two angles divided by $2\pi$ yields a number which we called the burial factor. A statistical study shows that the blocks associated with a fresh, sharply formed crater tend to be more emergent, more perched up on the surface than the blocks associated with

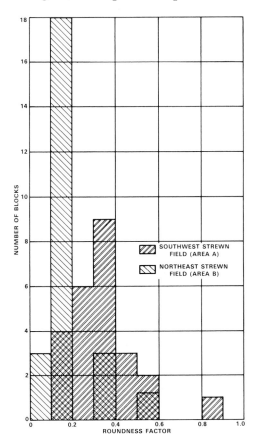

**Fig. 4-22. Distribution of roundness factors for blocks around a sharply formed crater.**

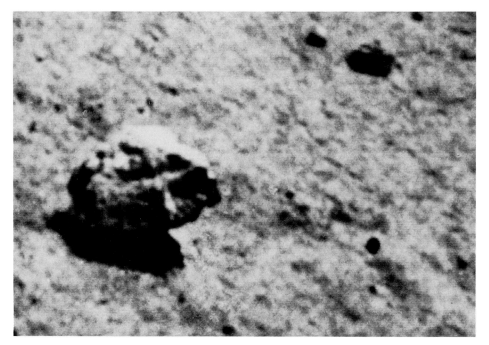

**Fig. 4-23. Some rocks lie on the surface like the one here which is about 15 cm across.**

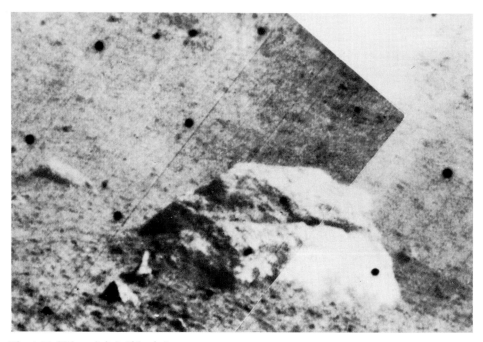

**Fig. 4-24. This rock is half-buried.**

Fig. 4-25. Some rocks have only a corner or edge protruding. Observations of many rocks such as shown in Figures 4-23, 4-24, and 4-25 led to the study and measurement of the burial factor.

Fig. 4-26. The first evidence of rocks with internal planar structure.

**Fig. 4-27. The landing site of Surveyor V in Mare Tranquillitatis.**

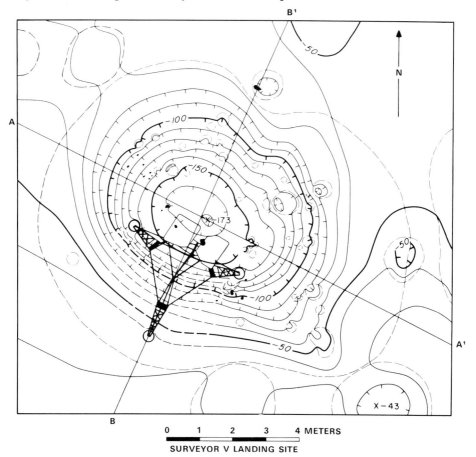

SURVEYOR V LANDING SITE

**Fig. 4-28. Topographic map of the crater in which Surveyor V landed.**

the subdued craters. So there are processes not only of grinding off the edges of a block, but also processes leading to progressive burial of these blocks over the course of time.

Some of the rocks which we saw inside the Surveyor III crater showed evidence of the internal planar structure. Some of them are markedly tabular, as in Figure 4-26; a close look reveals little sub-parallel ridges and grooves on the sides of the tabular block suggesting that there is an internal planar structure parallel with the large face on the block. From what is known now about the composition of the blocks, that Jim Patterson will tell you about shortly, I would guess that these blocks are probably flow-banded basalt.

SURVEYOR V

Surveyor V landed in Mare Tranquillitatis on the side of a small crater about 12 m long and 9 m wide. The spacecraft was tilted along the crater wall within a few degrees of the angle at which it might tumble. It was tilted about 20°, so the horizon has a high-amplitude sine wave on a cylindrical projection mosaic (Figure 4-27). The edge of the crater follows a line that roughly parallels the horizon. The black region in Figure 4-27 is a shadow within the crater formed by the edge of the crater to the left.

Figure 4-28 is a topographic map showing the form of the crater. The technique of focus ranging was used in constructing this map. Using the limited depth of focus in the 100 mm focal length mode, we were able to focus the camera in and out and take a series of pictures at a given azimuth and elevation and to range to various points on a lunar surface with fairly high precision. The contours are at 10 cm intervals. The map illustrates the elongate form of the crater. Actually the crater seems to be a double crater with two partly merged components separated by a low septum that comes out from the north wall. This double crater lies along a chain of craters which are parallel to the long axis of the double crater. It turns out that there are many similar craters on this part of Mare Tranquillitatis—either double craters or little chains of craters. They are all aligned in the same direction, very nearly northwest. This is the dominant structural direction on this part of the moon. Much larger linear topographic features on the border of Mare Tranquillitatis have this same northwest alignment.

An important feature of the Surveyor V crater is illustrated in the profile shown in Figure 4-29. The crater has no raised rim. The steepness of the walls increases somewhat in most places in the descent along the walls until a small, concave crater floor is reached. Craters like this were first seen in

**Fig. 4-29. Although steep-walled, this crater has no raised rim. Such "dimple craters" were first seen in Ranger pictures.**

**Fig. 4-30. The same crater as in Figure 4-29 photographed with a wide-angle lens at low-sun angle.**

the Ranger pictures. Take the little concave floor away, and just continue the walls downward a little, and one has what is called a "dimple" crater; dimple craters are commonly lined up in chains or rows. When they were first seen, it was inferred by Harold Urey that these craters were formed by drainage of fragmental material into a subsurface cavity. I think the explanation is entirely adequate for the crater in which the Surveyor V spacecraft landed. It was formed by drainage of the surficial debris layer of low cohesion into an underlying fissure which runs along the axis of the crater chain. Apparently, there are many such fissures cutting the bedrock of the *mare* in this region.

Figure 4-30 shows a chain of small craters lined up on the main crater wall. Probably they were formed by recent renewed drainage into the fissure. Another rimless crater, about 4 m in diameter, which is part of the crater chain can also be seen in the distance.

As the spacecraft landed on a slope, it slid down the crater wall, the footpads each leaving a trench about a meter long. This trench provides a good exposure of subsurface material in the debris layer as shown in Figure 4-31. Part of the floor of the trench has been smoothed by the sliding of the bottom of the footpad, and it appears bright at certain angles of the sun. In the 10-cm-high wall of the trench much of the material beneath the surface can be seen to consist of aggregates of one kind or another. Most are probably very loosely compacted aggregates derived by mild shock-compression of fine-grained material during repetitive cratering.

Some fragments that were kicked out by the sliding footpads cascaded down the wall of the crater and rolled out onto the floor of the crater beneath the Surveyor camera as illustrated in Figure 4-32. Many of them left tracks which have depths of 1, 2, or 3 mm. An upper bound for the strength of the upper few millimeters of the lunar surface here can readily be computed

Fig. 4-31. The subsurface material was exposed by the footpads of Surveyor V when it slid down hill.

**Fig. 4-32. Fragments kicked out by the footpads offer a clue to the strength of the upper few millimeters of the lunar surface.**

**Fig. 4-33. Fragments about 2 cm across show an aggregate character.**

**Fig. 4-34. Further evidence that some of the lumps are aggregates of smaller grains.**

**Fig. 4-35. A rock on the crater floor shows evidence that dark patches may be filled with fine-grain debris. The rock was photographed at intervals throughout the complete lunar day. This picture was taken at noon.**

if plausible densities for the fragments are assumed. It turns out that the strength can hardly exceed $10^4$ dynes/cm$^2$, which is very fragile indeed. One could easily poke a finger down into the surface for a few millimeters and hardly feel the resistance of the surface. The strength must increase rapidly with increasing depth because, within a few centimeters, it soon reaches a bearing capacity that will hold up the spacecraft, and would hold up a man. A man walking about on the surface would leave footprints approximately 2 cm deep.

Some of the fragments kicked out are clearly aggregates, as illustrated in Figure 4-33. One of the fragments, about 2 cm across, has little bright chips within it and other not-so-bright areas, which are surrounded by a darker matrix. Quite a number of fragments like this were seen. In the wall of the trench plowed by the footpad are objects which appear to be aggregates of aggregates (Figure 4-34). One dark lump about 4 cm in length, for example, is similar in its photometric characteristics to the fine-grained part of un-disturbed surface. I am reasonably sure that it is made up of fine-grained material, but it has distinct little bumps or warts sticking out on the surface, which are a few millimeters across, and these in turn are probably small, slightly compressed aggregates of fine-grained material; the whole lump is a lump of aggregates.

The aggregates of fine-grained material can be distinguished from rocks by the photometric function of their surface. In Figure 4-35 is a rock lying on the floor of the Surveyor V crater which was observed over much of the lunar day. The shadow of an omni-directional antenna marched right across it at noon, and the picture shown in Figure 4-35 was taken close to lunar noon. The exposed surfaces of the rock are very slightly brighter than the surrounding finer grained debris; there are some dark patches on the rock which are depressions filled in with darker fine-grain material. Figure 4-36 shows the same rock as it appears late in the lunar afternoon. Notice how much brighter the surface of the rock is relative to the dark fine-grained material. The photometric function of the rock surface is much more like that of a theo-retical Lambertian scatterer than is the fine-grained part of the lunar surface. The difference is due to porosity; the surrounding surface on the fine-grained material is very much more porous or open-textured than the rock surface, and I would infer that the rock itself has a much lower porosity than the fine-grained material. All of the larger angular fragments observed in the Surveyor picture not only are different in their photometric function but nearly all of them are also higher in albedo, commonly about 30 per cent higher.

Fig. 4-36. The same rock as in Figure 4-35 photographed late in the lunar afternoon.

Fig. 4-37. The brighter angular fragments have a photometric function different from those on the lunar surface. There is also visible a specimen formed by shock-melting.

In Figure 4-37 is a fragment about 5 cm across that may have been formed by shock-melting. Notice that various parts of the fragment are either much brighter or much darker than the surrounding fine-grained debris. This suggests that there is a specular component in the scattering function of this fragment, which in turn would suggest that it is partly glassy. The fragment has about the right shape to have been formed by shock-melting, as may be seen by comparing it with the objects shown in Figure 4-38. These are pieces, about the same size, of shock-melted sandy dolomite from Meteor Crater, Arizona. It has been something of a surprise to me that we have not seen many more objects like this on the moon.

## SURVEYOR VI

Surveyor VI landed in Sinus Medii very close to the middle of the subearth side of the moon (Figure 4-39) and close to a small but rather typical *mare* ridge (Figure 4-40). For the first time it was possible to get a good view from the lunar surface of one of these *mare* ridges, as shown in Figure 4-41.

**Fig. 4-38. Pieces of shock-melted sandy dolomite from a meteor crater in Arizona.**

**Fig. 4-39. Surveyor VI landed at this site in Sinus Medii.**

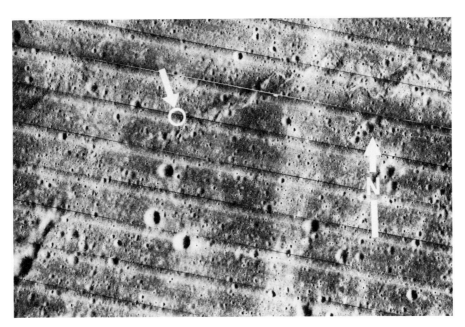

**Fig. 4-40. The open lunar crater in which Surveyor VI landed.**

Fig. 4-41. A lunar view of one of the mare ridges in the vicinity of the Surveyor VI site.

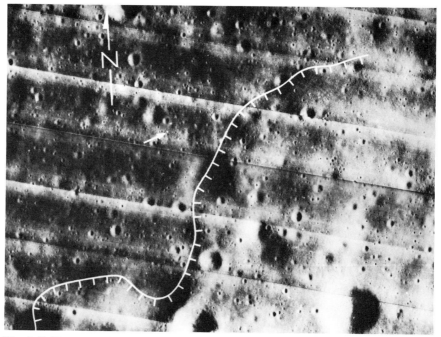

Fig. 4-42. Reconstruction of Surveyor VI site by triangulation. The white dot is the space-craft.

**Fig. 4-43. No trace of fragmental rims was visible from Surveyor VI.**

**Fig. 4-44. Photograph of crater with blocky rim taken by Surveyor I; compare this with similar features shown in Figure 4-43 taken from Surveyor VI.**

The foot of the *mare* ridge is visible and the crest of it is about 30 m higher than the adjacent plains area. Figure 4-42 shows a solution for the landing point, which was found by matching features that could be seen from the Surveyor camera and the features seen in Lunar Orbiter pictures; the little white dot is the size of the spacecraft.

A survey of the plains area revealed no trace within the line of sight of the Surveyor VI camera of a crater with a blocky rim (Figure 4-43). A number of craters with well-defined raised rims were seen, however, like the one shown in Figure 4-44, which is about 9 m across. The rim of this crater is smooth and devoid of blocks, unlike other craters of similar size seen at the Surveyor I site that have blocky rims. From the absence of blocks on the observable crater rims, the thickness of the regolith on the plains is estimated to be greater than 10 m.

Many coarse blocks were observed on the *mare* ridge, however, in strewn fields associated with craters. Figure 4-45 shows a crater about 30 m in diameter near the foot of the ridge, with many blocks on its rim. The debris layer on the ridge probably is about 8 to 10 m thick, significantly thinner than on the plains. This difference may be due to creep or flow of the debris from the ridge to the plains.

**Fig. 4-45. A 30-m-diameter crater with many blocks on the raised rim.**

Fig. 4-46. Debris thrown out by the footpad appears darker, contrary to expectations.

SURFACE ALTERATION OF THE REGOLITH

One of the surprises revealed in the Surveyor pictures is shown in Figure 4-46. This picture shows the debris thrown out by impact of one of the Surveyor I footpads. Notice that it is conspicuously darker than the undisturbed surface; its albedo is about 30 per cent lower. Everywhere that the lunar surface was disturbed either by a Surveyor footpad or by a mechanical claw carried on two of the successful Surveyors, or by rolling fragments, the surface turned out to be darker underneath. The surface of the moon as we see it optically in these pictures and at the telescope is photometrically different from the material underneath at depths no greater than a millimeter. The difference is opposite from what had been expected; subsurface material was expected to be lighter rather than darker than the surface material. Fresh craters about 10 m across or larger on the *mare* surface commonly have bright rims. The depth of these craters is probably a gauge of the depth of the debris layer. Fresh

**Fig. 4-47. Evidence from the debris layer in Nevada shown here suggests a model for the formation of the lunar debris layer.**

rock thrown out *is* brighter. However, in the fine-grained debris the immediate subsurface is dark.

For several years to come students of the moon probably will argue about why the regolith is darker than the subsurface. I prefer the hypothesis that there is some kind of coating deposited on the particles in the subsurface and that the coating makes them dark. The coating may be derived by sputtering; it may simply be an oxygen-depleted residue of sputtered material that is left on the grains in the subsurface. The coating is scrubbed away at the surface— it's completely scrubbed away on the surface of coarse fragments, and partially scrubbed away on the finer grained material.

MODEL OF DEBRIS FORMATION

The lunar debris layer or regolith is formed probably by repetitive impact— the same process that produces most of the small craters. The regolith looks somewhat like the layer of ejected debris formed around the nuclear crater Sedan (Figure 4-47). This crater was formed in rather typical desert alluvium composed of pieces broken up by various erosional processes, and the texture of the Sedan debris deposit is like that seen on the moon, because the crater was formed in material that was already fragmental. Most fragments seen in the debris layer on the moon are not produced by a single crater event, like Sedan, but represent the cumulative ejection from a great many craters.

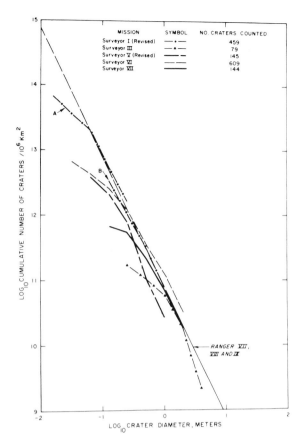

Fig. 4-48. The integrated size-frequency distribution of small craters also contributes to the model. There is a break at about 300 m in diameter.

Fig. 4-49. Example of the debris from a single crater in strong rock.

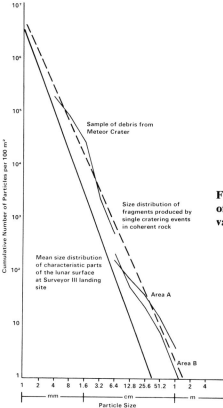

**Fig. 4-50. The size-frequency distribution of fragments in the strewn fields around various individual craters.**

The size-frequency distribution of small craters observed with the Surveyors is plotted on a log-log scale in Figure 4-48. This distribution is very nearly an extension to small sizes of the average size-frequency distribution of craters first observed for the lunar plains areas with the Ranger spacecraft. Below crater diameters of 100 m, the slope is −2, the cumulative number of craters is inversely proportional to the square of the crater diameters. Above 100-m crater diameter the slope is steeper. There is good reason to believe that the size distribution of craters smaller than 100 m is essentially a steady state distribution of craters, that is, as many craters are being eroded away in a period of time as are being formed. Thus the surface is being plowed and turned over and over by small craters. I think the average *mare* surface has been turned over and re-cratered approximately three hundred times by craters of 1-m diameter and larger.

Around a crater formed in strong rock one finds the kind of debris shown in Figure 4-49. This debris has a characteristic size-frequency distribution which differs from that observed in the lunar debris layer, but is similar to the size-distribution of fragments in the strewn fields around individual craters

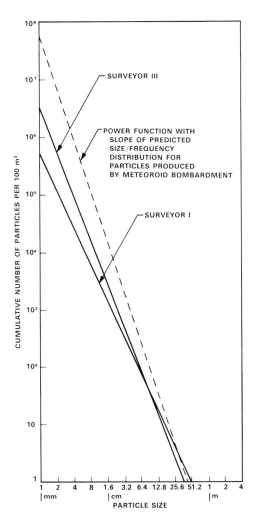

CUMULATIVE NUMBER OF PARTICLES PER 100 m²

10⁸
10⁷
10⁶
10⁵
10⁴
10³
10²
10
1

SURVEYOR III

POWER FUNCTION WITH
SLOPE OF PREDICTED
SIZE/FREQUENCY
DISTRIBUTION FOR
PARTICLES PRODUCED
BY METEOROID BOMBARDMENT

SURVEYOR I

1    2    4    8   1.6  3.2  6.4  12.8  25.6  51.2  1    2    4
| mm                | cm                          | m
PARTICLE SIZE

**Fig. 4-51. Slopes of size-frequency distributions of fragments from the different Surveyors are different.**

as shown in Figure 4-50. For comparison, the average frequency distribution of the fragments in the debris layer at the Surveyor III landing site is shown. There is a considerable separation between these two distributions. Figure 4-51 shows the variation in the size-frequency distribution of fragments seen on the *maria*. The slope of the Surveyor III distribution is the steepest; at the Surveyor I site the slope is gentlest. The steepness of the distribution function is correlated with the thickness of the debris layer and the crater density on the *mare* surface. The size-frequency distribution of fragments in a young, thin debris layer is initially like that produced by single cratering events. With the passage of time and continued cratering, particularly with small impacting objects, the mean size of the debris gets finer and the slope of the size-frequency distribution function becomes gentler and gentler. The thickness of the debris layer simply corresponds to the depth to which the debris layer has been stirred by cratering just once. The calculation of the thickness of the debris based on the inferred numbers of craters that have been formed corresponds quite well to the thickness determined from the depth of the smallest blocky-rim craters.

DISCUSSION

(Q) As the Surveyor set down on the lunar surface did you notice any dust clouds raised by the impact?

(A) Shoemaker: Unfortunately, high-speed television picture sequences would be necessary to see such a cloud. The spacecraft lands blind, and then after a considerable period after landing the cameras are turned on so that any fragments that might be kicked up would have either gone way out of range or would have settled back down again. However, there is very good evidence, particularly from Surveyor III which landed several times, during which the camera mirror was open and pointed in the direction to get the first picture, that a considerable amount of fine-grain material was sprayed up from the surface. The material either pitted or was coated on the exposed part of the mirror, and we observed strong veiling glare in the pictures. A similar effect was observed on Surveyor I but to a limited degree. After those experiences, we finally persuaded the Hughes Aircraft engineers to shut the mirror down and close the camera before landing, so the mirror would be clean. We have seen good evidence that a spray of very fine-grained material did come up from the surface.

(Q) Could that be the cause of the different reflective values of the dark vs light material? The very fine material had been removed and the reflectance value would be reduced.

(A) Shoemaker: When we first saw the Surveyor I pictures, we considered that as one of the possibilities. It turns out that the surface right around the spacecraft, as measured from the Surveyor pictures themselves, has the same albedo as the surface farther away, and it has the same albedo as that which we can measure for a still broader area at the telescope. In fact, in some cases, the local albedo is lower because we're in a little dark patch. So the surfaces we see are not just coated with fine, bright material locally; we are looking at a surface that probably has had some disturbance but not very much. Also the photometric characteristics right in the immediate vicinity of the spacecraft seem to be representative of a much larger region of the lunar surface. The differences in albedo cannot be attributed to the disturbance upon landing.

(Q) Would temperature changes account for the erosion effects?

(A) Shoemaker: The erosion of the blocks appears to be due primarily to the bombardment by small particles, both interplanetary meteorites as well as little secondary fragments of the moon itself. Probably it is the secondary fragments that would produce the differential erosive effects, because high-speed particles won't tend to erode differentially whereas low-speed impact particles will. There may be same contribution from evaporation of material, either by sputtering or by ultraviolet radiation. We need to do more laboratory studies, I think, on that process before we can really evaluate to what extent it might contribute to the loss of material from the lunar surface. Probably the impact process is very nearly adequate if not entirely competent to produce the erosive effects that we see.

(Q) In the pictures showing the rocks embedded in the surface, how do you tell whether a given rock has been worn down more or is embedded deeper?

(A) Shoemaker: We looked for correlation between roundness and burial and found none. In other words, there are angular blocks that are embedded as well as rounded blocks. In fact, we did such a study with the statistics which I showed and found that there is a lack of correlation between degree of rounding and burial. But there is a general broad correlation such that, on the average, a fresh crater will have blocks that are more emergent from the surface as well as more angular.

(Q) Most of your pictures of the lunar surface were illuminated by low-angle illumination. Were there any provisions for making stereophotographs of the surfaces?

(A) Shoemaker: This is one of our big disappointments in the Surveyor project. Our original experiment had two cameras on the spacecraft, but only one camera was ever flown. On Surveyor VI we finally tried and successfully completed the experiment to move the spacecraft from one position to another on the moon. About the middle of the lunar day the Surveyor VI spacecraft engines were fired up briefly. It was launched from the lunar surface and landed again about 8 ft away, which was just about the distance we had hoped to use for stereo base. Thus we do have stereoscopic pictures from Surveyor VI. Unfortunately, the pictures that can be matched and studied in stereo are taken under high sun so that, while we can easily recognize rock fragments and carry out the photogrammetry from this experiment, they're not very dramatic pictures when observed stereoscopically. They do provide the opportunity to measure closely the shapes of things on the surface. We had to resort to subterfuges to get the geometry in the other missions.

(Q) Have you ever found bedrock in place?

(A) Shoemaker: The examples closest to bedrock, or to a piece that is anchored to the rest of the moon, probably were observed by Surveyor VII. I simply didn't have time to discuss Surveyor VII today, but Jim Patterson will tell you about some of the chemical results. There are large blocks sticking up from what seem to be flows; I think they are flows of hot debris on the flank of Tycho, and these blocks may very well be anchored in the flows. We did not see any large rocky fragments that seemed to be anchored to the rest of the moon on the *maria*. In fact, there are very few places on the *maria* where one might look for anything like that. One could hope to find bedrock in the upper parts of the walls of some very fresh craters, but it would be inside the crater, so that we didn't land in the right places to see them.

(Q) You referred to a process called sputtering. What is it?

(A) Shoemaker: It is the evaporation of material along the entry path of a relatively low-energy solar proton. If the material is evaporated away into free space, it will just leave the immediate vicinity, but if it is evaporated in the pore spaces beneath the surface much of it will be redeposited on the surfaces of the grains. It is a fairly well-known effect in metals, for example. Some experiments have been done on rock powders, but I think the results are inconclusive.

(Q) What is the explanation of the craters found in straight lines?

(A) Shoemaker: I think the craters that are lined up are those not formed by impact. The impact craters always have a raised rim when they are fresh. Even though it doesn't appear as such, most of those little craters do have a raised rim. This is one of the things we could see much better in the stereo pictures than in the monoscopic views. But the craters that are lined up are rimless; they are just depressions in the surface. They appear to be due to drainage of the debris into a subsurface fissure; their formation is controlled by a subsurface structure. There are other rows of craters on the moon which are due to a pattern of bombardment of secondary fragments. When we look at the larger patterns of secondary craters on the moon we very often find beautiful little chains of craters, formed by a related cluster of fragments thrown out together from the main crater. I brought none of those pictures with me.

# Chemical Analysis of the Lunar Surface JAMES H. PATTERSON

*Abstract: Exploiting the backscattering of alpha particles and proton production by alphas, three devices have been placed upon the lunar surface successfully by Surveyors V, VI and VII. The results of each exploration are not inconsistent and suggest a basaltic type of material. Details of the experiments are presented and the interpretation of the data is discussed.*

## INTRODUCTION

The chemical analysis of the lunar surface was carried out on the same series of spacecraft which Dr. Shoemaker has described. The last three of these spacecrafts, Surveyors V, VI, and VII, carried the alpha-scattering device which was used to perform this chemical analysis. The head of the team of scientists which conducted the alpha-scattering experiment is Anthony Turkevich of the University of Chicago. Also on the team with Professor Turkevich and me is Ernest Franzgrote of the Jet Propulsion Laboratory.

## DESCRIPTION OF THE APPARATUS

The method of analysis used in this experiment is different from the usual methods of chemical analysis—one that has not been used heretofore. Thus it merits a description. The principle upon which it is based was discovered sixty years ago in the studies of the properties of alpha particles by Rutherford and others. They found that alpha particles were scattered in a backward direction, which was very surprising at the time because such behavior was contrary to the theory of the structure of matter then in use. Such backscattering was one of the facts which led Dr. Rutherford to his famous nuclear theory of the atom. Previously, it had been thought that mass and charge were distributed rather continuously throughout the atom, but from the backwards scattering of alpha particles Rutherford was able to show that the mass and the positive charge in atoms were concentrated in an exceedingly small volume and the alpha particles were scattered backward from these high-density

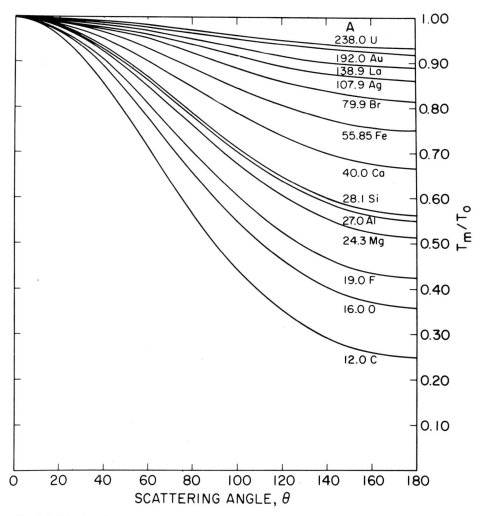

**Fig. 5-1. The fractional energy retained by the alpha particle as a function of its scattering from various target nuclei.**

nuclei, as he called them. From this discovery has developed the nuclear theory. Another phenomenon used in the alpha scattering method of analysis is the production of protons by the interaction of alpha particles with nuclei of light elements. For this bit of physics we are again indebted to Ernest Rutherford. It was the first artificial transmutation, which he carried out around 1920 using alpha particles on nitrogen to form $O^{18}$. Both of these processes have been known for many years, but because of technological difficulties it was not feasible to use them for analytical purposes until recently.

The principal energy relationships for these interactions were proposed as long ago as 1911 by Darwin, who used the formula

$$T = T_o \frac{(X \cos\theta + \sqrt{1 - x^2\sin^2\theta})^2}{(1 + X)^2} \qquad \text{Equation 1}$$

T is the energy that remains with the alpha particle after it has collided with a nucleus heavier than itself; $T_0$ is the initial energy of the alpha particle before the collision; X is a ratio of the mass of the alpha particle to the mass of the nucleus with which it has collided, and $\theta$ is the angle of scattering of the alpha particle. So you'll notice that the energy of the alpha particle immediately after the collision is dependent only upon the energy before scattering, the angle of scattering, and the ratio of the masses. From this, it can be seen that by measuring the energy of the alpha particles scattered at a known angle, the mass of the nucleus which scattered the alpha can be determined.

Figure 5-1 is a plot of Equation 1 for various atomic masses. Two conclusions can be drawn from this plot: (1) At low atomic masses there is the greatest resolution. The mass increment from 12 to 16 shows a very large separation in energy ratios, but from 192 for gold up to 238 for uranium there is a very small difference in energy of the alpha particles scattered at a given angle. Thus the method has much better resolution for elements of lower atomic mass. (2) The resolution is much better at large scattering angles. At 0° scattering angle, of course, the resolution is zero, but the maximum resolution occurs at 180°, though from 160° to 180° the change with angle is very small. This is fortunate because one has to use a range of angles. In our experiment we used an average angle of 174 ± 3°, so we exploit the high resolution and the small change with angle.

We calibrated our apparatus by using pure elements as scatterers. Figure 5-2 shows results for five elements as obtained with an early version of the device. Platinum, silver, and chromium have essentially flat plateaus, each ending with a very sharp break. Boron, on the other hand, has structure caused by the alpha-proton reaction that occurs in several of the light elements in

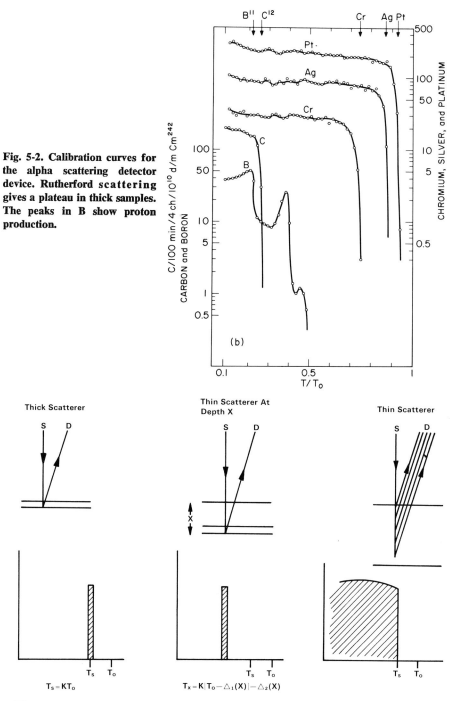

**Fig. 5-2. Calibration curves for the alpha scattering detector device. Rutherford scattering gives a plateau in thick samples. The peaks in B show proton production.**

**Fig. 5-3. The observed plateau is an integration of scattered alpha energies that have been displaced due to energy loss from interaction with electrons in a thick sample.**

the periodic table. Thus we observe both Rutherford scattering and proton production.

The existence of the plateau effect shown in Figure 5-2 can be explained with the aid of Figure 5-3. If the sample is considerably thinner than the range of an alpha particle in the material, a very narrow peak is obtained because of the scattering relationship in Equation 1. However, alpha particles that interact with nuclei lower than the top layer of the material have some energy lost through the interaction with electrons within the material so that, when they reach the point where they are scattered, they are already degraded in energy; when they return through the material to be detected they lose some more energy. Thus the unique alpha energy from a material of a given depth is displaced from the monoenergetic value by the amount that is lost in entering and returning, so that instead of a sharp peak, a plateau is obtained.

The distribution of alpha backscattering intensities as a function of atomic number is shown in Figure 5-4 on a log-log scale. The solid points were obtained with elementary samples. The open circles are derived points obtained on compounds. For atomic numbers higher than about 12 the data fall on a straight line with a slope of 3/2. This agrees with the theory which predicts a 3/2 power for alpha elastic scatterings. Below atomic numbers of approximately 10 or 12 there is an enhancement of the intensity from the scattering process which is considered to be due to resonance scattering, in which an alpha particle either enters the nucleus or at least becomes intimately associated with the nucleus, and an alpha particle is then re-emitted, not strictly from elastic scattering. Due to the three-halves power law, the intensity

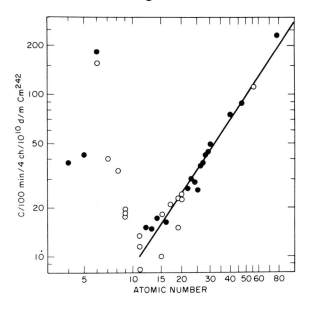

Fig. 5-4. Calibration curve for the Surveyor alpha scattering package. The intensities are plotted against atomic number.

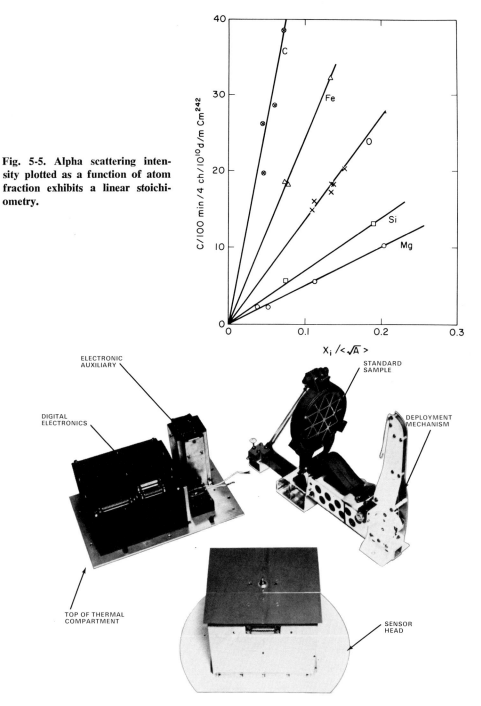

**Fig. 5-5. Alpha scattering intensity plotted as a function of atom fraction exhibits a linear stoichiometry.**

**Fig. 5-6. The basic units of alpha backscatter and proton production instrumentation for the Surveyor missions.**

becomes very low for atomic numbers of less than 10 so that although the resolution would be high the intensity would be low for low atomic number elements; but, fortunately, there is quite high intensity for such elements as carbon, oxygen, nitrogen, boron and beryllium. Especially for oxygen, which is of considerable interest, this is very fortunate for our method of analysis.

We now look at the stoichiometry of the method. Figure 5-5 shows the plot of the atom fractions divided by a constant, which is characteristic of the sample, plotted against the intensities. It is approximately linear for alpha scattering by each of the elements shown. The same can be done for protons, and similar results are obtained.

In the Surveyor program the experimental instrument consists of the basic units shown in Figure 5-6: the sensor head, which was deployed to the surface of the moon for the actual measurements; the digital electronics and the electronic auxiliary, which were stored in a thermal compartment of the spacecraft; and the deployment mechanism, which was the device for storing the sensor head during transit and deploying it to the surface of the moon. Figure 5-7 is a drawing of the sensor head. There are six alpha sources, which provide a collimated beam of alpha particles that strike the lunar material through an opening in the bottom of the box. The particles are scattered back to two alpha detectors in a narrow cone at a fixed angle of 174°. There are four larger proton detectors, which detect and measure the energy of the protons that are obtained from the nuclear interactions. The proton detectors also have asso-

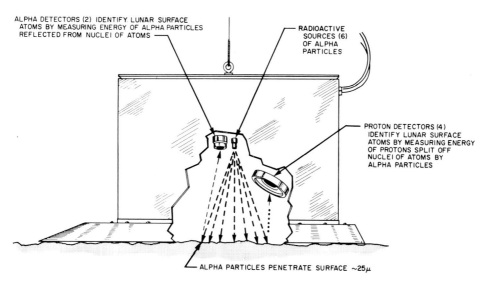

Fig. 5-7. The sensor head which was deployed to the lunar surface after the Surveyor landed on the moon.

ciated with them guard detectors, other proton detectors in anticoincidence with the proton detectors to provide cosmic ray shielding. The alpha detectors are gold-coated surface-barrier silicon detectors; the proton detectors are lithium-drifted surface-barrier silicon detectors with a gold sheet of about half-mil thickness to absorb the alpha particles so that they will not be detected. The bottom opening of the sensor head is shown in Figure 5-8; this is the moon's-eye view of the sensor head. The six alpha sources, which are of curium-242, are shown in an array in the center. Very closely associated with them are the two alpha detectors, and around the sides are the four larger proton detectors.

LUNAR OPERATION

Figure 5-9 is a diagram of the procedure for operation on the moon. The sensor head is stowed in the position shown on a platform in the spacecraft; also on this platform is a standard sample which has been analyzed on earth

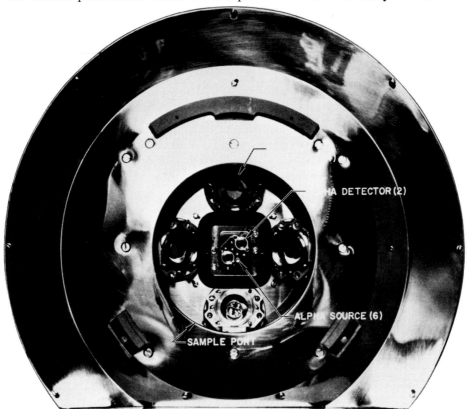

**Fig. 5-8. The moon's-eye view of the sensor head; the six alpha sources, two alpha detectors, and four proton detectors can be seen through the sample port.**

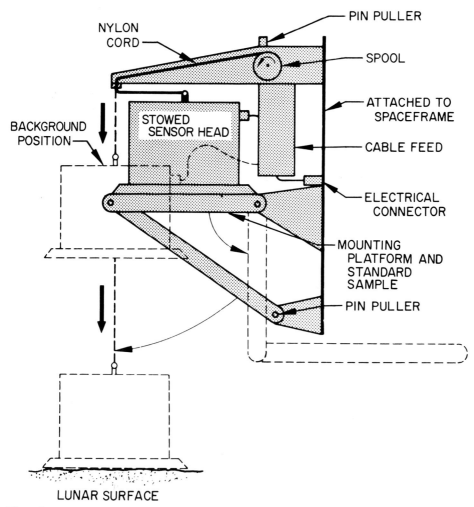

**Fig. 5-9. The sensor head deployment operational procedure on the lunar surface.**

both by standard chemical means and by the alpha scattering method. It is again analyzed after the moon is reached in order to calibrate the instrument and to verify that it is operating correctly. After enough data have been obtained in this phase, a squib is blown and the platform falls away leaving the instrument suspended from a nylon string approximately 56 cm above the surface of the moon. Since in this position the scattering from the moon is very slight, these data are used to obtain a background reading arising from cosmic rays or from any general radiation that may be present on or near the surface of the moon. After operating in this mode for a few hours, another squib is blown, and the sensor head is let down gently to the surface of the moon. This was the proposed normal operation. It did indeed function properly on the

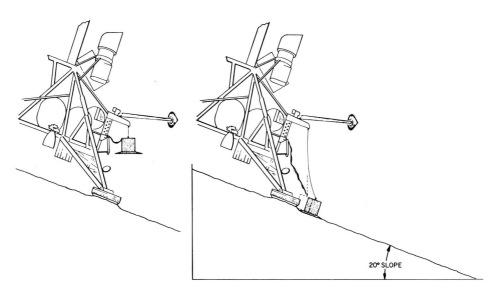

**Fig. 5-10. On Surveyor V the sensor head was placed beyond its normal location due to the slope of the landing site.**

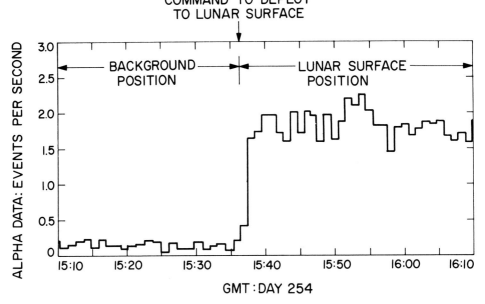

**Fig. 5-11. Typical increase in intensity accompanying deployment to the lunar surface.**

first two flights but on the third, which was Surveyor VII, something went wrong and it stayed in the background position when the second squib was blown. Fortunately, the surface sampler (the "digger" as it is commonly called) was also aboard this mission and it was used to push the sensor head down to the surface of the moon so that analysis could be done after all. If the malfunction had occurred on the other two missions no analysis could have been made.

In Figure 5-10 can be seen the position of the sensor head relative to the spacecraft on Surveyor V. It was at a 20° slope on the side of a crater, so that the sensor head was deployed at a greater distance from the spacecraft than it would have been if on a flat lunar surface. Dr. Shoemaker described the trench that was dug out by the footpad of the spacecraft as it slid down the side of the crater. Some of that trench material was thrown out to the place where the alpha scattering device was later laid down. Hence, material was analyzed that was not necessarily surface material but could have been as much as 10 cm below the surface. After two days of operations, the spacecraft rocket engines were fired, not enough to move the craft but enough to move the device from the dotted position to the solid position. Thus there were two samples about 4 in. apart on this first mission on the moon. They were separated by just about the width of the window. In every case, the sensor head was deployed when the moon was not visible to the receiving station at Goldstone, California, so that we could not get television pictures directly to verify deployment to the surface of the moon. There always was a typical dramatic increase in intensity at the time that it was deployed, as shown in Figure 5-11. In this way we knew that on Surveyor VII the sensor head had not been deployed, because the same low intensity continued after the squib was blown.

Through the courtesy of Dr. Shoemaker, television pictures of the instrument in place on the surface of the moon were obtained. Part (a) of Figure

**Fig. 5-12. Two views of the sensor head resting on the lunar surface at two different positions, before and after firing of the vernier rocket.**

5-12 shows the position of the first analysis of the moon and part (b) the second position after it had been moved slightly by firing the spacecraft engines. The deployment mechanism and the string which lowers the sensor

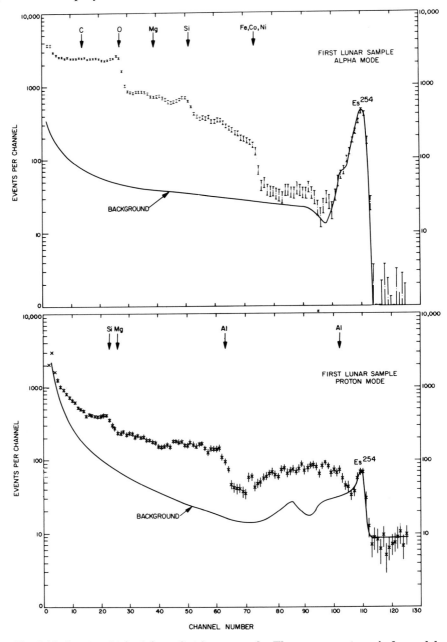

**Fig. 5-13. Spectra obtained from first lunar sample. The upper spectrum is from alpha particle scattering and the lower spectrum is proton production.**

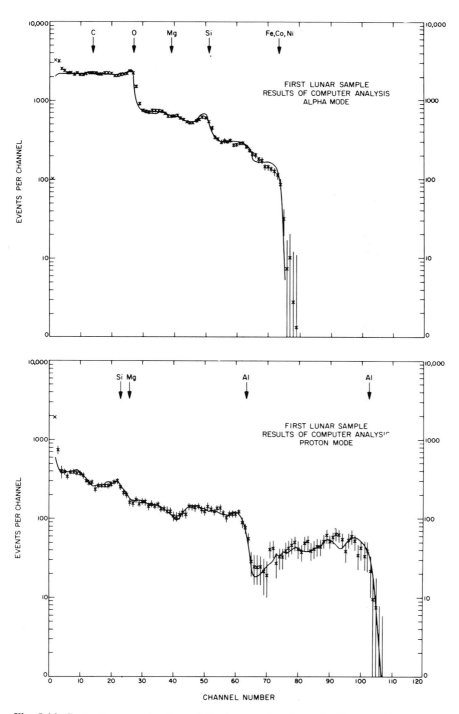

**Fig. 5-14. Computer reconstruction of the two spectra shown in Figure 5-13. The solid line is a least squares fit.**

can be seen. Also visible is the cable that carries the power to the sensor head and the signals back to the spacecraft.

## RESULTS

The spectrum obtained from the first sample before the engines were fired is shown in Figure 5-13. The alpha spectrum in the top half of the figure shows the principal features of the analysis—a pronounced break for oxygen, a slight break for magnesium, a very noticeable break for silicon, and the iron group shows a strong break. The small break at the right in the region of gold does not suggest gold on the moon; it is gold inside of the instrument. Although the sources were well enough collimated so that they did not see inside of the instrument (they saw only the window) we did not allow for the aggregate recoil that deposited some radioactive material on a film placed over the end of the collimators; this material "saw" the inner sides of the sensor head and gave us a gold break. So don't start a gold rush to the moon. The solid line is the measured background as smoothed out for plotting. The peak on the right is from a small amount of einsteinium placed on the detectors as an energy marker. The lower half of the figure is the proton spectrum with the background and the einsteinium peak. The most significant features of the proton spectrum are these: the two aluminum peaks and weaker breaks due to silicon and magnesium. The other features are less pronounced and had to be resolved by computer.

Figure 5-14 shows the least squares computer fit to the background-less effective spectrum of both the alpha particles and the protons. The solid line is the fit by the computer. The computer solves 180 simultaneous equations, one for each of the energy channels of the proton and alpha particle spectrum combined, to obtain the best fit. The equations were of the form

$$KI = \sum_i X_i^o \, I_i^o \, \sqrt{A_i} \qquad\qquad \text{Equation 2}$$

where I is the intensity in a given channel, K is an energy independent constant characteristic of the sample, $X_i$ is the unknown atomic fraction of the element i, $I_i^o$ is the intensity of the element in its elementary form, and A is the atomic weight.

Each pulse received by the detectors is sent back individually to the earth from the moon by telemetry, with an energy code on it. The primary data to be used to get the best final analysis are recorded on magnetic tape. These tapes contain the complete record of all data pulses, but for use in the missions and for immediately after the missions there was also a system of accumulating pulses in spectra, like the data in Figure 5-13. This information was sent by teletype in the form of a read-out of the counts per channel for 128-channel analyzers for the proton and alpha spectra. This was used during the mission to monitor the operation and to get a quick look at the results. Through the services of computers at the Jet Propulsion Laboratory in Pasadena, eight elements —carbon, oxygen, sodium, magnesium, silicon, calcium, aluminum and iron —were fitted to the curve.

The present preliminary analysis is merely the best result obtainable from this crude analysis. That is why there are such high limits of errors on the analysis—because final results have not yet been received. Although it is a rather crude method of doing it, with only eight elements, one can see that there are no large fractions of other elements, or else breaks in this curve would be observed that haven't been accounted for. Notice that the computer curve of the alpha spectrum fits fairly well except at the place where one expects calcium. The spectrum for the moon may have a small amount of some other component which masks the sharp break for the calcium. For this reason on the Surveyor V mission, we did not give calcium and iron groups as separate results, and quoted only the total of calcium and iron. We do not know what this means yet. Carbon, oxygen, sodium, magnesium, aluminum and silicon are normally well resolved from each other. With calcium, the resolution becomes rather uncertain, and it is not possible to resolve iron from nickel, cobalt or from chromium and magnesium, so that to say iron really includes the other elements in the region of iron. Likewise, calcium also includes potassium. We hope that we shall be able to get some resolution between calcium and potassium in our final results, and there is a very, very faint possibility of identifying nickel separately if nickel is a very large fraction of the so-called iron.

The computer breakdowns of the spectra obtained on the Surveyor V are illustrated in Figure 5-15. The upper solid line is the computer fit. The other lines are the spectra used by the computer to make up the solid line. Carbon is weak here; in fact, it hardly appears at all. The oxygen is this plateau that accounts for the oxygen break, and similarly for the other elements. Only four elements of the eight appear in the proton spectrum. Aluminum is the most prominent; it goes clear out to the end. Sodium is the next high-energy one. Magnesium and silicon are at lower energy. These are the actual curves that gave the best fit of the total spectrum as seen by the computer at this time.

It is revealing to compare the results from Surveyor V with some typical rock and meteoritic samples as shown by the bar graphs in Figure 5-16. The heavy bar on the left of each element is the result obtained with the alpha

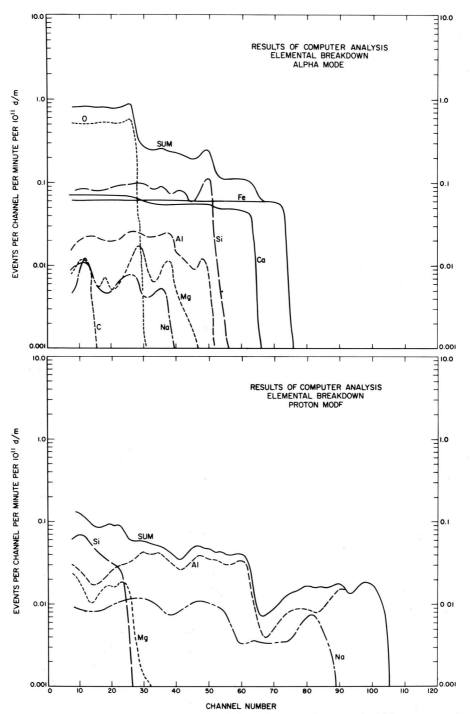

Fig. 5-15. Computer analysis for an input of eight elements, the sum of which are constrained to give the least squares fits in Figure 5-14.

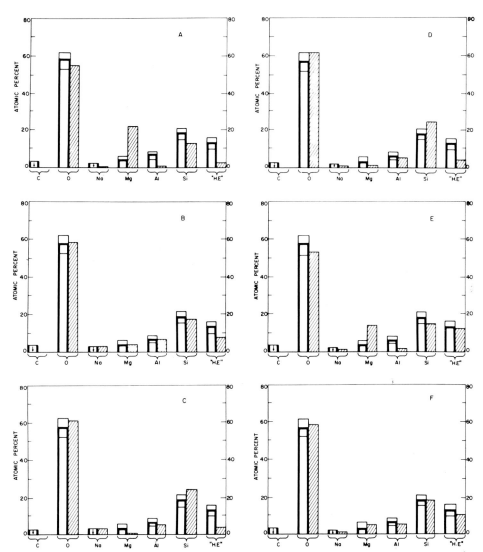

Fig. 5-16. Comparison of Surveyor V eight-element analysis with compositions of known samples. Ca is included in the Fe group under the "H.E." designation. The rock samples are: A – dunite; B – basalt; C – granite; D – tektite; E – chondritic meteorite; F – basaltic achondrite.

scattering method. The limits of error are shown with lighter lines. The shaded bar next to it is the result for a typical rock. Part A is dunite. In this case the magnesium that was found was much too low, far outside the limit of error for magnesium. The aluminum found was much too high. Silicon was slightly high also and heavy elements, which include calcium and iron and elements of that region, are much higher. Part B is a comparison with a typical basalt. The fit is quite good except for the heavy elements which are quite variable anyway, but the basalt we have has a lower heavy-element content of calcium and iron than the Surveyor V. Part C is granite; our magnesium is high in this case instead of low as it was for the dunite. The aluminum fit is not too bad, but our heavy elements are much higher than ordinarily found with granite. Similar results are obtained with a tektite shown in part D. In part E is shown the comparison with a chondritic meteorite which is considered by some actually to come from the moon, so that it was predicted that it would be the composition of the moon. But it quite obviously does not fit because the magnesium in the chondrite is much higher than that found on the moon and the aluminum is much lower. The basaltic achondrite in part F is another form of meteorite, and it fits much better—even better than basalt because of the heavy elements. All the others are within the limits of error, including the heavy elements.

The results of the mission on Surveyor VI were very similar to those of mission V. Dr. Shoemaker has described Surveyor VI, and on this mission data were obtained during the first half of the lunar day. When the spacecraft hopped to one side the instrument was turned upside down facing the sky and could no longer obtain data, but enough data were obtained before that to get an analysis. Surveyor VII was a mission to the highland area just north of the crater Tycho on an area considered to be part of the material ejected from the crater when it was formed. On this spacecraft were magnets and the surface sampler as well as television and the alpha scattering device. One magnet was on a footpad. The surface sampler (or "digger") also had a magnet installed in the hope of picking up iron samples.

The surface sampler provided for analyses at three different points as shown in Figure 5-17. Superimposed on the drawing of the spacecraft is the pie-shaped sector which represents the range of the surface sampler's reach. It gave access to the spot no. 1, quite close to the position which the sensor would have had under normal deployment. At spot no. 2 was a rock, roughly 7 cm in diameter, and the sensor was put directly on top of that rock for analysis. At position no. 3 the digger plowed up the surface of the moon to reveal some subsurface material, for a very short period of analysis. Unfortunately, in all three of these positions the distances of the sample from the detectors and sources were different, because in the first position some small rocks held the instrument up away from the surface of the moon. Of course, in the second position this rock was small enough to protrude inside the window of the instrument. In the third position it was put down right over a hollow where material was dug out and one could not see the depth of the hollowed-out material. Since it was not in a standard position the analysis is a

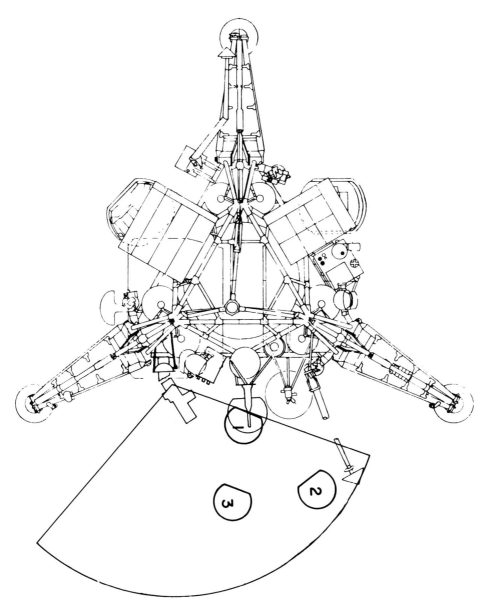

**Fig. 5-17. From Surveyor VII the lunar surface was sampled at three points.**

bit more complex but not impossible. There is a tentative analysis for sample no. 1 at this time.

Figure 5-18 is a picture showing the position of sample no. 1 after the sensor head was moved to sample no. 2. The position of the slight depression from the rim of the device can be seen, and the circle drawn in is the actual sample area that was measured at this time. The rocks that prevented the sensor from lying flat on the lunar surface caused a departure from standard geometry. A valid analysis will come out of it after some further experimentation with geometry. Figure 5-19 shows the position of the rock which protruded into the window. The circle is the position which the window occupied when the sensor head was in place. The outer rim can again be seen in another place where it was accidentally put down too soon. The arm that carries the surface sampler out is clearly shown although the sampler itself is not visible. We can also see the edge of the alpha scattering instrument sitting in position no. 3. Figure 5-20 shows the third position before the sensor head was placed there, and again the arm of the surface sampler. There is a shadowed spot right under the position of the sampler where the surface sampler has been digging. The final position of the alpha scattering instrument when the sun

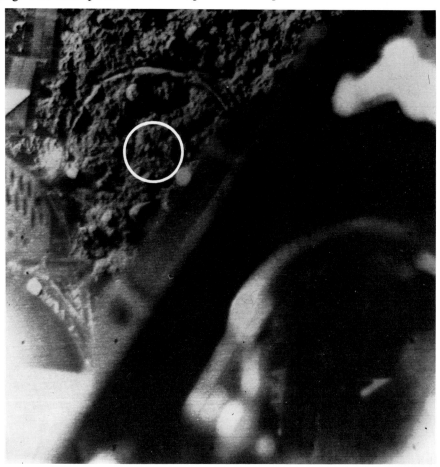

**Fig. 5-18. Close-up of the site of sample no. 1 after the sensor was moved to position no. 2. The circle shows the position of the sample port during the analysis of sample no. 1.**

went down (position no. 3) is shown by the white lines drawn on the photograph. Operating again on the second lunar day delivered more data in this position. Unfortunately, the proton detector of the system had broken down partially during the very cold lunar night so that only the total number of proton counts per minute could be obtained—no spectral measurement. The alpha system was good so that it was possible to resolve the iron and calcium, using this later data.

SUMMARY AND CONCLUSIONS

A comparison of the alpha spectra of the three different samples from Surveyor V, VI, and VII is shown in Figure 5-21. The solid line is from Surveyor

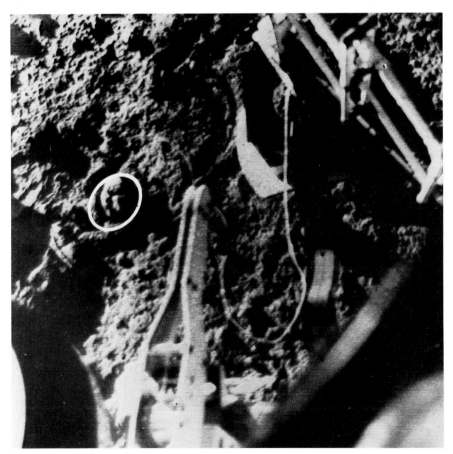

Fig. 5-19. At position no. 2 a small rock protruded into the sample port, the position of which is shown by the circle. The sensor head was being moved to position no. 3 when this photograph was taken.

**Fig. 5-20. A photograph of position no. 3 before the sensor head was placed. The superimposed drawing shows the sensor head location to scale.**

V, the dotted line for Surveyor VI, and the individual points with the error bars are from Surveyor VII. Notice that the lines for the Surveyor V and VI were very close—the lines deviate only slightly, except at the point where in Surveyor V we do not resolve the calcium and the iron. We did resolve them in Surveyor VI, so we were able to quote analyses for iron and calcium separately in that system. Results from Surveyor VII are significantly different from the other two, the main difference being that the iron was less than half of the concentration of the iron in the other two Surveyors.

A summary of the results for the three Surveyors as of the present time is shown in Table 12. Essentially no carbon was found. Our upper limits are 3 per cent, 2 per cent and 2 per cent for the three samples. Oxygen percentage

**Table 12. Chemical Composition of the Lunar Surface at Surveyor V, VI and VII Sites (Preliminary Results)**

| | Percent of Atoms | | |
| --- | --- | --- | --- |
| Element | Surveyor V | Surveyor VI | Surveyor VII |
| | | | (Sample I) |
| Carbon .................. | < 3 | < 2 | < 2 |
| Oxygen .................. | 58 ± 5 | 57 ± 5 | 58 ± 5 |
| Sodium .................. | < 2 | < 2 | < 3 |
| Magnesium .............. | 3 ± 3 | 3 ± 3 | 4 ± 3 |
| Aluminum .............. | 6.5 ± 2 | 6.5 ± 2 | 8 ± 3 |
| Silicon .................. | 18.5 ± 3 | 22 ± 4 | 18 ± 4 |
| "Calcium" .............. } | 13 ± 3 | 6 ± 2 | 6 ± 2 |
| "Iron" .................. } | | 5 ± 2 | 2 ± 1 |

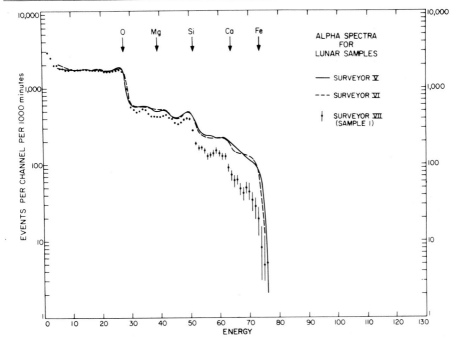

**Fig. 5-21. Comparison among alpha spectra for three different lunar samples. Note sample no. 1 on Surveyor VII is quite different from those on Surveyors V and VI.**

was about the same—58, 57, 58; no difference within the limit of error. Again with sodium, only an upper limit could be set. There is certain to be some sodium present; a definite number should be available for it later when the complete analysis is made. Magnesium was about the same in the first two samples but was slightly higher in the third. The limits on the first two do not exclude zero, but this is not the same as saying it is less than 6 per cent

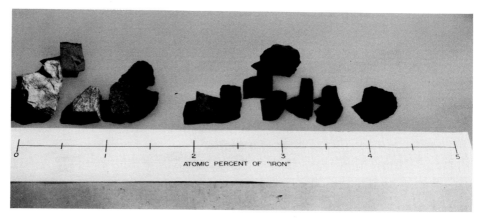

**Fig. 5-22. Possible dependence of albedo upon atomic per cent of iron. (After Franzgrote.)**

because it is thought that 3 per cent is a good value for it; on the Surveyor VII it was $4 \pm 3$ per cent. Aluminum was higher in Surveyor VII than in the other two. Silicon in Surveyor VI for some reason came out higher. This may be due to an interaction between the oxygen and the silicon in the alpha spectrum which caused it to be somewhat higher. The calcium and iron on Surveyor V can be given at the present time only as a combination because of lack of resolution. On Surveyor VI and VII calcium and iron could be shown separately. The striking difference is that of only 2 per cent iron in the highland region near Tycho compared with 5 per cent in Surveyor VI and roughly the same in Surveyor V. The results of Surveyors V and VI gave very similar compositions. The only difference that might appear to be significant is silicon, but it is not believed to be a true difference. Surveyor VII has possibly higher magnesium, probably higher aluminum, and definitely lower iron than the other two.

Franzgrote has made a study of the albedo for different samples of material which have different atomic per cent of iron (Figure 5-22). Such studies may possibly offer an explanation of why the highlands have a higher albedo than the *maria*. The points in the highlands for analyses had a lower percentage of iron than the points seen in the maria. Since only three sites were visited, one can make no strong claim for its validity, but this is a possible explanation.

DISCUSSION

(Q) I am intrigued by the difference between computed spectrum of Ca and Fe and the observed spectrum. Of course, one of the important rock-forming minerals is titanium. Can you detect it?

(A) Patterson: This might account for the lack of definition between calcium and iron. Titanium is the one element in the best position to be observed if there was something else between calcium and iron because vanadium, chromium, and manganese are too close to iron in atomic weight, but titanium is right in between. Although we tried to use our crude data to analyze for titanium we found nothing conclusive. Of course, we got a positive result but it is not really very definite yet. We hope when we finally analyze these data to have a definite number for titanium. There could well be some titanium there.

(Q) I noticed a considerable fluctuation in your data over a period of one lunar day. Did you include these in your error bars?

(A) Patterson: Oh, yes. The event rate is rather low, a few per second for the alphas and about one per second for the protons; this gives a fluctuation with time. In this analysis each one of these points represents seventeen hours (1,000 minutes) of operation and we have summed the data over all this time. There were quite large fluctuations within individual batches, but we depend on the statistics of summation of long time operation to obtain our analysis. Actually the statistical errors are less than the ones we have quoted. We have added further error because of possible systematic error. We have temperature-corrected these data, but a close look at the spectrum shows the fit at the sharp breaks such as oxygen is not perfect. Evidently there is some energy correction yet to be done. The statistical error obtained from the count rates are smaller than the errors we have quoted. We hope to improve our results by more detailed analysis of the primary data which is more extensive in total number of events. We can also correct for such things as noise because it's time-coded. We hope eventually to come out with a much better analysis including several more elements and a much better limit of error. Also we expect to do some experimentation with the geometries that we used on the last three samples from Surveyor VII to try to improve the analysis.

(Q) Wouldn't a better analysis be obtained by waiting until a sample is brought back from the moon?

(A) Patterson: Yes. This *was* the last Surveyor—the last mission before man goes to the moon. The next step in the moon program, of course, is to bring samples back, and they will be analyzed much more extensively than we have done. We have only done an analysis of the major constituents, but lunar samples in earth laboratories will be analyzed for minor constituents and many other things. This present work is important because it is a first look at the moon. On the other hand, Surveyor VII may be the only analysis we have of that type of material for a long time, because, as Dr. Shoemaker mentioned, the astronauts in the Apollo program will be landed in an equatorial belt, and I doubt that they will venture up into the highland areas very soon.

# Luminescence and Chemical Effects of Solar Protons Incident upon the Lunar Surface

EDWARD ZELLER

*Abstract: Materials similar to those present on the lunar surface have been bombarded by protons under conditions analogous to those found on the moon. Sputtering and hydration account for the observed production of OH, and probable formation of CH and SiH. As a control OD was also produced. During the time scale for proton interactions with lunar material production of water and organic compounds is likely. Both the luminescence and the optical characteristics of the moon can be profoundly affected by proton-induced chemical changes within the upper few microns of the lunar surface.*

INTRODUCTION

My co-workers in this research are: Paul Levy at Brookhaven National Laboratory; Luciano Ronca, formerly of the U.S. Air Force Cambridge Research Laboratories, now with Boeing; and Miss Gisela Dreschhoff who is currently working with me. Both the AEC and the U.S. Air Force have aided in supporting this research.

LOCALIZED LUMINESCENCE

There are two general types of lunar luminescence phenomena, those which are localized and those which cover the entire lunar surface. The localized phenomena are almost always associated with young lunar craters. There are three general types of localized phenomena. First, there is the very strong red luminescence which has been reported by Kopal near the crater Kepler and by Greenacre near the crater Aristarchus. In these cases the luminosity is so high that it is visible on the sunlit lunar surface. Therefore it is very doubtful that it is any kind of solid-state luminescence process because the luminosity is too high.

Another group of occurrences that may fall into the category of luminescence phenomena relate to the fact that occasionally some of the lunar craters appear to be filled with gray clouds. These gray clouds are quite conspicuous and appear almost as though they are luminescent. They have been observed many times over long periods, particularly at times of lunar perigee. At such times the gravitational field on the moon is strongest and could cause openings of fissures with small gas eruptions which may account for the gray clouds which fill craters.

Finally, there is another group of localized phenomena which appears as the simultaneous luminescence of whole crater areas such as the craters Kepler and Aristarchus. The entire crater area will simultaneously light up and glow for periods up to thirty minutes. The light is so intense that spectral measurements have been made. Unfortunately, the spectra are not consistent and their quality is not sufficient to give good information about the elemental abundances. We do not yet know whether the material is gaseous.

GENERAL LUMINESCENCE

The luminescence from the entire surface of the moon can be classified into two types: (1) luminescence of the sunlit surface itself, and (2) luminescence during eclipses.

Because of the vast difference in intensity the luminescence of the sunlit surface is not directly visible. Its presence is revealed by measurements of the spectra of sunlight reflected from the moon. The Fraunhofer lines in the reflected spectrum are different from those observed in direct sunlight. Some radiation appears at the wave lengths of the Fraunhofer lines. This has been considered as evidence that there is luminescence taking place during the time that the sun is actually shining on the lunar surface, an interpretation which is open to argument.

There is another type of luminescence which is much more conspicuous and can be seen by direct observation. Some of you may have seen it. During some total lunar eclipses we are still aware that we can see the entire lunar surface. The moon does not completely vanish during these total eclipses. In other cases, the moon does vanish so completely that we cannot see it even with telescopic observation. Vaucouleurs, a French scientist, showed that the

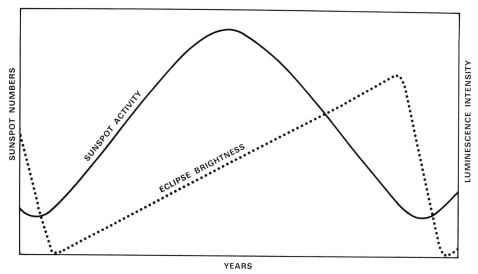

**Fig. 6-1. During total eclipses of the moon the degree of darkness of the lunar disc is related to the sunspot cycle. Eclipses are darkest shortly after the sunspot minimum, and maximum eclipse brightness occurs about one year before sunspot minimum. (After Vaucouleurs.)**

degree of darkness of the eclipse was related to the sunspot cycle. His results are summarized in Figure 6-1 which shows that the eclipses are darkest shortly after the sunspot minimum, and middle brightness is reached at sunspot maximum. The reasons for the four-year phase lag between the sunspot maximum and the maximum eclipse brightness are definitely not yet clear. Vaucouleurs showed that this relation has existed over a number of sunspot activity cycles which have been folded together in the figure.

Hopmann has suggested that the eclipse luminescence is related to the position of the sunspots relative to the lunar equator. At maximum he argues that the sunspots occur in the middle latitudes, and at minimum the sunspots tend to occur near the lunar equator. He points out that because the sunspots do occur near the lunar equator most of the radiation is emitted in the plane of the ecliptic and therefore more of it strikes the moon at that time. This is true for photon radiation such as x-rays, ultraviolet, and gamma rays. However, the charged particles are deflected by the solar magnetic field and therefore they are channeled into the plane of the ecliptic. Although ultraviolet and gamma rays are not deflected during the eclipse, they do not impinge upon the lunar surface; they are screened by the earth just as is visible light.

PREVIOUS EXPLANATIONS

Many explanatons have been put forward to account for the various kinds of luminescence.

THERMOLUMINESCENCE

It has been postulated that there is a sudden increase in brightness of the lunar surface just as it heats up when the sun first strikes it at lunar dawn. Considerable effort has been made to detect this flare. It has been argued that it is an example of thermoluminescence. This is certainly a possible mechanism for the sunlit moon, but it cannot account for the eclipse luminescence for the simple reason that during an eclipse the lunar surface temperature falls rapidly and thermoluminescence cannot take place.

Photoluminescence has been suggested and the process could work very well for the sunlit moon. Optical photoluminescence results from the optical bleaching of trapped charges, and that mechanism is probably present on the sunlit moon. Obviously there is no sunlight to cause the bleaching in the case of the eclipsed moon.

Direct excitation by incident radiation is certainly one of the most logical explanations. This is clearly possible for the sunlit moon. Protons, electrons, and photon radiation could serve to excite directly the sunlit surface. Photons are screened from the eclipsed moon by the earth, but charged particles may not be screened. The geomagnetic tail of the earth may focus the charged particles on the moon at the time of the eclipse.

Why then is there not maximum eclipse brightness at the time of maximum sunspot activity? One major group of luminescence phenomena that has been ignored as a possible explanation for the lunar luminescence is chemi-luminescence. To invoke this as an explanation of phase lag one has to argue that there are very large-scale chemical changes on the lunar surface which are not instantaneous. If, for example, any kinds of free radicals or meta-stable species are present within the uppermost portion of the lunar surface, substantial quantities of energy might be stored within this zone. Processes can be postulated which could permit the release of this energy as visible light. Some process has to be found for very large-scale chemical alteration of the lunar surface as well as the large-scale production of free radicals and unstable compounds or meta-stable species. It is quite clear that the uppermost atomic layers of the lunar surface have the principal influence upon the luminescence and optical properties of the moon. When one looks at the moon from the earth one sees effectively only the upper few microns, but really very little of the moon's bulk. Therefore, very little of its mass would be involved to change its optical and luminescence properties enormously. The proton reactions described below could cause substantial alteration of the lunar surface chemistry and might produce some type of chemi-luminescence.

PROTON EFFECTS

It is known that the proton flux incident upon the moon can and probably does cause major alterations in the composition of the uppermost hundred microns of the lunar material. In order to evaluate the possible chemical reactions which might take place, we must examine in some detail both the projectiles (protons) and the target (lunar material). There are three groups of protons having widely different energies which are incident upon the lunar surface.

The solar wind protons have energies of a few kiloelectron volts with an average flux of about 4 x $10^8$ particles cm$^{-2}$ sec$^{-1}$, or a flux of $10^{16}$ particles cm$^{-2}$yr$^{-1}$. The solar flare protons range above 1 MeV in energy, and have a flux between $10^{11}$ and $10^{12}$ particles cm$^{-2}$ yr$^{-1}$. The galactic background of the true cosmic rays is made up of about $10^8$ particles cm$^{-2}$ yr$^{-1}$ with energies exceeding 1 GeV. The important feature is the variation in the flare proton flux with the variation in sunspot cycle ranging from $10^9$ to $10^{12}$. Table 13 shows the details.

The lunar surface rocks are probably silicates with compositions well within the range of terrestrial rock. Normal silicates can be considered to be essentially packing structures of oxygen ions, and it is upon this target that the incoming solar and cosmic protons impinge. Since the kinetic energy of the proton determines the range of possible interactions with the target, the three energy groups will have decidedly different effects upon the lunar rocks. The very high energy protons can penetrate the surface quite deeply and they can produce nuclear reactions. It is easy for a GeV particle to penetrate a meter or more, and then to engage in a nuclear reaction at that depth. As a result of the

**Table 13. Variation in the Solar Proton Flux Occurs with the Variation in the Sunspot Cycle. The Average Flux is About $10^{11}$cm$^{-2}$yr$^{-1}$.**

| | Protons Incident on the Moon | | |
|---|---|---|---|
| Energy Range | Flux, Particles/cm$^2$ sec | Solar Activity | Reference |
| Few KeV | 4 x $10^8$ | Quiescent | White (1965) |
| ~1 MeV | $10^2$ | Quiescent | |
| >10 MeV | 350 | Flare | |
| >20 MeV | 250 | Flare | Paulikas et al. (1966) |
| >40 MeV | 100 | Flare | |
| >80 MeV | 5 | Flare | |
| ~1 BeV | Occasionally | Flare | Singer (1959) |
| >10 MeV | $10^2$ | Avg. over last 10 yrs. | Lal and Venkatavaradan (1966) |
| ~0.5 Mev | $10^3$ | | Estimated from above |

nuclear reaction between 1 and 10 protons of about 1 MeV will be produced for every GeV proton. Thus, nuclear reactions will not be absolutely confined to the extreme surface of the lunar material.

The solar protons with energy of the order of 1 MeV will penetrate the lunar surface between 10 and 100 $\mu$. Most of their energy is lost through ionization, thermal excitation, elastic and inelastic scattering. As long as the particle possesses more kinetic energy than the binding energy for any potential chemical bond that it might form, it cannot react with the lattice chemically. When the kinetic energy has dropped to the range of chemical bond energy, that is, between 1 and 20 eV, it may be able to react chemically with one of the lattice atoms, or it could capture an electron and simply become an hydrogen atom. Unreacted hydrogen atoms are not very likely to remain in the lattice for long periods because of the high degree of ionization and thermal spiking which the particles produce in their path through the lattice. Thus it is very unlikely that any of the hydrogen which might be produced in this fashion will remain uncombined.

The solar wind protons penetrate the surface less than 1 $\mu$, but they can also react chemically. Even though these reactions do occur, it is highly unlikely that any of the reaction products will remain very long, because they will be sputtered off the surface almost as rapidly as they are formed.

## OH AND OD PRODUCTION

In 1965 we started experimental work at Brookhaven National Laboratory. We chose to use the artificial equivalents of flare protons in order to produce the products deep enough within the target that they would not be lost to the surface by diffusion or by sputtering. We bombarded glass targets with 1-MeV protons from a Van de Graaff accelerator. The targets showed a progressive increase in the OH absorption band at 2.8 $\mu$ in the infrared. In Figure 6-2 the heavy line is the original absorption; there was a small amount of OH present which appears as an absorption line. After we irradiated the glass with increasing proton doses the absorption increased. There is no doubt that substantial quantities of hydroxyl ions are produced in silicates, and that they will tolerate a considerable degree of heating. Some of the glass targets were deformed by heat. Calculations indicate that somewhere between 10 and 100 per cent of the incident protons reacted to produce OH. Similar experiments have shown that hydroxyl ions are produced by bombardment of calcite and synthetic rutile.

In order to confirm the reaction we irradiated a quartz plate with deuterons. A very strong OH peak was present in the original plate. The results are shown

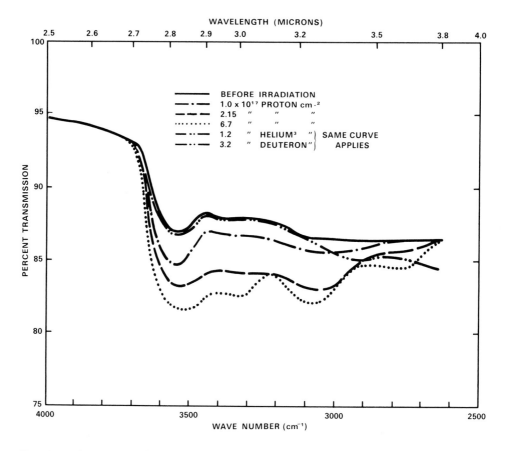

**Fig. 6-2. Infrared absorption spectra of a silicate glass plate before and after the indicated radiations.**

in Figure 6-3. The solid line is the original, the dashed one is the irradiated material, and the differential curve is derived. Because deuterium has a greater mass than hydrogen, we expect the peak to be shifted deeper into the infrared. The absorption does occur at exactly the predicted location. There seems to be no question that we can produce OH or OD by bombarding silicates and other oxygen-rich substances with protons or of deuterons with energies of the order of 1 MeV. The absorption band was very well developed, and we found that subsequent irradiations increased the intensity of the band.

LUNAR PROCESSES

What is the significance of these reactions in relation to the lunar surface? The sun is losing hydrogen at the rate of several millions tons per second. The

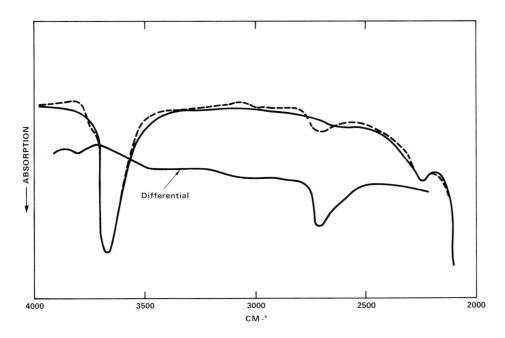

**Fig. 6-3. Infrared absorption spectra of a quartz plate irradiated with deuterons. The spectrum before irradiation showing a strong OH peak is represented by the solid line. After irradiation the dashed line shows the expected isotope shift deeper into the infrared.**

amount which strikes the moon every year is only about 5,000 tons. After about a thousand years this would be enough to cover the entire lunar surface with water to the depth of 1 $\mu$, if all the protons which reacted to produce the water were captured and if none of it were sputtered away into space. The protons which penetrate the lunar surface more than 1 or 2 $\mu$ are much less abundant, but within a few million years they, too, would produce significant effects.

Figure 6-4 represents a possible theoretical model of the surface of the moon. The log of the depth of ionic layers is a silicate lattice is shown as the ordinate. One assumes that the surface is continually sputtered away by the solar wind and strong reduction occurs because of the effect of selective oxygen sputtering. Immediately beneath this reduction zone is a zone of relatively high water concentration. So much OH is produced here that the possibility of having another proton enter the lattice and stop within the range necessary to form a water molecule becomes a reasonable probability. Beneath that is the zone of OH production, which might produce some of the coloration effects discussed earlier. Eventually the limit of charged particle defect production is reached. This is followed by a region of defect production by uncharged

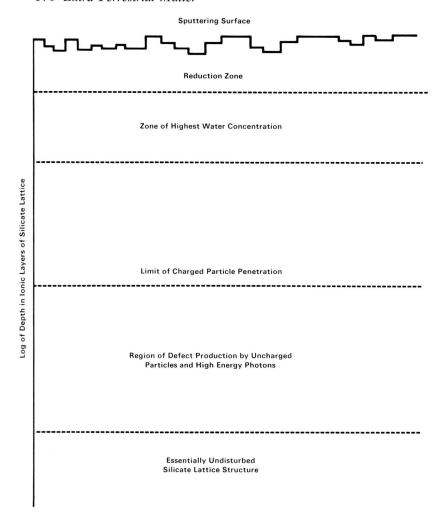

**Fig. 6-4. A model of the lunar surface showing the depth of the zones at which various processes reach maximum yield.**

particles and high energy photons. If any hydrogen is produced and it diffuses into this region, it can of course react. Finally, the essentially undisturbed silicates would appear at still greater depth. Surface churning could bring the hydrated surface rock to some considerable depth below the lunar surface. The moon may thus accumulate a substantial amount of hydrated material in the upper part of the regolith.

Radar reflectivity measurements made by Hagfors suggest that a tenuous surface layer exists on the moon. This could, in part, be the result of chemical reactions. It is also observed that the tenuous surface layer is more conspicuous in the older areas of the moon than around young craters. Perhaps the tenuous surface layer observed by radar represents the results of chemical alteration by proton bombardment. The term "space weathering" has been used to describe these effects.

Thus far we have considered only reactions between protons and the oxygen atoms present in the silicates.

However, any element which is capable of forming a compound with hydrogen should be able to react when it is irradiated with high-energy proton beams. To test this, carbon in the form of diamond was bombarded with protons and deuterons ranging in energy from 0.7 to 1.5 MeV. Diamonds were chosen because of their transparency to infrared, which permits evaluation of the reaction products in place. The irradiations were performed at liquid nitrogen temperature. Figure 6-5 is a photograph of the beam tube of the Van de Graaff. Visible are the viewing port used to check the target alignment and the dewar which contains the liquid nitrogen. The targets are held on a rod which is in contact with liquid nitrogen. During the irradiation, a large amount of luminescence appears. In this case the beam is of the order of about $10^{13}$ particles $sec^{-1}$ on calcite.

The results of these experiments are much more complex and difficult to evaluate than those in which OH or OD is produced because there are large

**Fig. 6-5. The Van de Graaff beam tube showing the target inspection part and the liquid nitrogen dewar.**

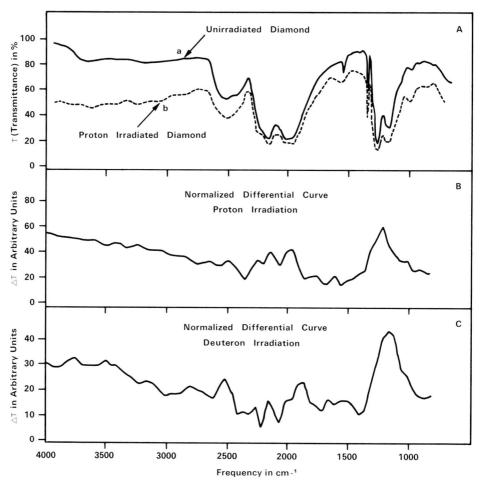

**Fig. 6-6. Infrared absorption spectra of diamond irradiated by protons and deuterons at liquid nitrogen temperature.**

numbers of variations. The formation of blebs and layers of graphite by the thermal spike effects cause additional difficulty in the interpretation of the diamond data. Some of the fundamental vibrational modes for unsaturated hydrocarbons are definitely present and clearly visible. Part A of Figure 6-6 shows the data for the irradiated diamond. One can see the infrared absorption curve (a) for the diamond target before irradiation. Irradiation increased the absorption greatly, as shown by the dotted curve (b). The normalized irradiated diamond curve subtracted from the normalized unirradiated curve are shown in B. Small blebs, or layers of graphite, cause a substantial amount of scattering. One of the most conspicuous absorption peaks is obscured by the scattering. It is clear that several peaks arise from hydrocarbons. By irradiating with deuterons it should be possible to shift those peaks to longer wavelengths. Note that the intensities are not to be compared directly because the scales are different. It appears possible to produce hydrocarbons by 1-MeV proton bombardment of diamond. Silicon carbide crystals were also irradiated, and bands for both CH and SiH may have been produced.

In the diamond experiment the intensities of some of the infrared absorption bands were found to change with time after the irradiation. This suggests some kind of reorganization of the reaction products. Our current investigations show that some bands increase in intensity while others show simultaneous decreases. This suggests a complex chemical reorganization process. Currently, we hope to determine whether the shifts in the peaks and the changes in bands are accompanied by the emission of light. The fact that proton-irradiated quartz shows its peak thermoluminescence emission at 320°C. while the same sample irradiated with gamma rays shows its peak at 260°C., with no trace of the 320°C., peak suggests that the processes differ in their basic mechanisms.

The clay mineral dickite is known to luminesce very strongly after it has been heated to drive out part of the water of hydration. Upon cooling, the mineral rehydrates rapidly from the air and shows a very strong luminescence associated with this process. The phenomenon, discovered in Switzerland by Grögler and Stauffer, is the only example known to me of a geological material which luminesces strongly with falling temperature. Obviously, this reaction with this specific mineral does not occur upon the lunar surface; but other related reactions may take place, for there certainly is sufficient hydroxyl ion present. Perhaps there is luminescence resulting from the hydration or rehydration of minerals which occurs while the surface is relatively cold during the eclipse. Indeed, the "mineralogy" of the lunar surface is likely to differ greatly from that of the underlying silicate, which should have rather familiar mineralogy. The "minerals" of the lunar surface should not be expected to be in stoichiometric equilibrium. The lunar mineralogist is likely to find that he is much more interested in radiation vs. temperature than the more familiar type of pressure vs. temperature relationships which concern terrestrial mineralogists.

The so-called minerals present in the upper few microns of the lunar surface will tend to be in equilibrium with their total radiation environment and with their electrical charge environment. There may be migration of charged ions through the surface material if the charge structure on the lunar surface is strong enough. Alterations in the radiation and electrical charge environment could bring about reactions which could cause light to be emitted. The surface equilibria may undergo progressive changes with the variation in the radiation flux which is related to the sunspot cycle. Although the solar wind is fairly constant, the cyclical variations of the flare proton flux could have marked effects upon reactions near the surface. Finally, charged ion diffusion in solids is strongly affected by the accumulation of electrically charged zones, either directly at the surface or within the solids. This could cause relatively rapid migration of hydroxyl ions, for example. If the surface is charged positively, they'll be drawn toward the surface, and conversely a negative surface charge would push them deeper in the lunar material.

CONCLUSIONS

Both the luminescence and the optical characteristics of the moon can be profoundly affected by chemical changes within the upper few microns. The principal chemical effects of the protons are twofold: (1) sputtering with selective removal of the oxygen and the attendant reduction of the metal ions as well as the formation of hydrides, and (2) hydration of the silicates directly below the reducing zone. Experiments have shown, then, that OH and CH and possibly SiH can be formed by proton bombardment in vacuum, and it appears likely that other so-called organic molecular fragments including NH can be formed. If the surface composition is favorable, organic molecules could be produced on the lunar surface, asteroidal surfaces, the surfaces of meteorites or comets, or on any solid bodies exposed to the protons. These need not be massive reactions involving vast tonnages of material. They are confined to the surface, but they are continuous. Finally, production of water and organic compounds in cosmic dust grains will be very much favored by the large area which is exposed to the protons, and any accumulation of these grains on the lunar surface could also affect both its luminescence properties and its optical characteristics.

DISCUSSION

(Q) Isn't the lack of total darkness of the lunar disc during total eclipse usually explained by refraction of sunlight around the earth due to its atmosphere?
(A) Zeller: It has been suggested, but there is no universal agreement about it.
(Q) Is it possible to use your method of quartz irradiation for a dosimeter?
(A) Zeller: Yes, there is a rather extensive program on the use of various solids of that sort and thermoluminescence of solid materials in dosimeters for protons, gammas, neutrons, etc. Thermoluminescent dosimeters are commercially available.

(Q) I particularly want to separate neutrons from gammas simultaneously.

(A) Zeller: I think you could do it.

(Q) Is it possible that the quartz did not have uniform lattice purity throughout? The surface traps might be different from the bulk.

(A) Zeller: That is possible, but we used exactly the same plates. We heated them, then irradiated them, heated them, got the glow curve. We then irradiated identical pieces with cobalt gamma rays and ran them over again by thermoluminescence methods. The protons do not penetrate the entire depth of the plate, whereas the gammas do penetrate; there is a bulk effect. When we irradiated with gammas we found that the quartz turned a dark, smoky color. This was not present in the proton-irradiated material, probably because the protons penetrate only a few microns.

(Q) If water is formed in the upper layers of the moon, would you expect it to stay there or would it escape to the vacuum of outer space?

(A) Zeller: I'd expect it to stay there for quite a while. When I first started this study I thought it would leave, but we found that we could irradiate glass in vacuum until it actually began to soften and still retain the OH in it. I would certainly have expected the OH to be driven out in the vacuum of the Van de Graaff, but it was not. I am convinced now that we get such a firm bond that the OH will not be lost. On the moon, of course, the temperature is far below that of our glass sample. In fact, the whole process is favored by a low temperature. That is one of the important factors in the production of water or organics with the gas phase reactions. In these cases the reaction is favored by high temperature which can't be maintained very long in the solar nebula. The proton process is one that goes on continuously in space even at a few degrees Kelvin. The same process should occur any other place where cosmic dust and proton plasmas are found. We need not confine it to the moon or our solar system.

(Q) What is the temperature of the surface of the moon in full sunlight?

(A) Zeller: It varies considerably but its average is about $400°$ K. This is not nearly enough to drive off the OH.

(Q) My question refers to your first slide. Many years ago I taught in an astronomy course that the main reason for the different brightnesses of the lunar eclipses was that the moon's orbit is elliptical and hence at different distances the moon enjoys a different fraction of the earth's umbra. Was this effect taken into account in the curves of your first slide?

(A) Zeller: I've heard this explanation myself. My first slide showed the work of Hopmann (at the observatory at the University of Vienna) and Vaucouleurs who have done a correlation with the sunspot cycle. Frankly, I am not in a position to evaluate the difference.

(Q) It would be a little risky since the moon's orbit does not have an eleven-year period.

(A) Zeller: True, and in this sense it should not fit the sunspot cycle curve.

(Q) I am disturbed by the lack of data points on the curve.

(A) Zeller: The original papers do have the points on them. This was a curve I made to illustrate phase lag. Obviously, the only real way to solve this problem is to make photometric measurements of the lunar surface during eclipses.

(Q) It is reasonably certain that a major part of the illumination of the surface of the moon during a lunar eclipse is due to light refracted by the earth's atmosphere. This is a consequence of an experiment done by Surveyor V where we had the spacecraft look back at the earth during the eclipse and see the ring of light around the earth being reflected. We also observed interruptions of the ring by meterological phenomena in the upper atmosphere which depended upon how far down into the atmosphere the light passed through. We did not look back at the moon to see if the light came from a rather small source or some more diffuse source such as the solar corona.

(A) Zeller: This may be the answer to the problem. But I still don't understand why it would appear that the darkest eclipse time would be just after the sunspot minimum and the brightest would be just before. I don't see why that should be true.

(Q) Were you able to make a numerical justification of the rather high intensities of the luminescence phenomena seen against a high solar illumination?

(A) Zeller: No. I don't believe we can do that with any solid material now known. I don't know any solid phospher that one could stimulate directly by ionizing radiation which would give this kind of effect. This is one of the reasons why I believe the phenomena reported by Kopal and Greenacre may be gas eruptions, lava, or something of that sort. I find it very difficult to believe that these are solid-state processes.

(Q) Kopal's work is based upon only one observation. The contrail of a plane could have passed in front of the moon. Greenacre's report, on the other hand, is based upon very careful painstaking work.

(A) Zeller: I don't think there is very much question that we are faced with something that is real, and I quite definitely feel we cannot explain it. It does not appear to be a solid-state phenomenon.

# III. Cosmic Rays

# The Chemical Composition of Galactic and Solar Cosmic Rays

## FRANK B. McDONALD

*Abstract: Recent developments in balloonborne and spaceborne dE/dx detectors, Cerenkov counters, and emulsions, and in laboratory fossil track studies, along with better measurements of fragmentation parameters, have yielded new data on stripped cosmic ray nuclei from Z=1 to Z=92. Relative nuclear abundances seem to differ significantly from the universal cosmic abundances. It is now possible to separate solar cosmic rays from galactic cosmic rays. There appears to be no unique accelerating mechanism, but instead several mechanisms seem to make significant contributions. Although the details are not yet understood, these mechanisms must account for a total cosmic ray energy of $10^{56}$ ergs and a supply rate of some $10^{42}$ ergs $sec^{-1}$.*

INTRODUCTION

The term cosmic rays does not describe a unique phenomenon but includes many different channels through which information is obtained about the cosmos in its broadest sense. Among these are: (1) starlight, (2) x-rays, (3) $\gamma$-rays, (4) 3° K. black-body radiation, (5) radio waves, (6) classical charged particles (including electrons), (7) neutrinos. The classical definition of astronomy—something that one looks at through a telescope at night —is broadened considerably by these new channels. I'll try to show that conventional cosmic rays, that is, charged particles, have essentially moved or been pushed from the realm of high-energy physics into the realm of astrophysics. In effect, this is a reversal of the former order of events. Historically, high-energy physicists studied the interaction products of the cosmic ray flux from the standpoint of high-energy particles. In this way they found the positron, the muon, the charged pions, the neutral pions, the heavy mesons, and the strange particles. Now they are reversing the procedure and starting to do the beam studies after they have studied the interaction products.

To a large extent, the studies of the primaries rest on two basic technological developments. One of these is the improved polyethylene Skyhook balloons, and the other (not necessarily in order of importance) is the space program. These two developments have been of great importance in enabling scientists to get above an appreciable amount of the atmosphere to look at the beam before it has interacted.

A large-scale view of the cosmic rays starts with the flux of protons. At the top of the atmosphere are found approximately 0.2 protons $cm^{-2}$ $sec^{-1}$ $ster^{-1}$, a relatively low flux. The flux of alpha particles is approximately 0.03 $cm^{-2}$ $sec^{-1}$ $ster^{-1}$, and the flux for $Z>2$ is approximately 0.003 particles $cm^{-2}$ $sec^{-1}$ $ster^{-1}$. The total nucleon flux is approximately 0.38 nucleons $cm^{-2}$ $sec^{-1}$ $ster^{-1}$. These are the gross features of a low-intensity beam made up of roughly 85 per cent protons, 13 or 14 per cent alpha particles, and 1 per cent heavier nuclei. Here a basic problem is already encountered. The universal abundance as determined from meteorites and other sources predicts that the heavier nuclei should constitute less than 1 per cent of the alpha particle flux instead of the 10 per cent actually measured. It is difficult to explain why there are so many heavy nuclei and why the flux of electrons is only approximately 0.01 times that of the protons. In any case, we shall see that the alpha particles and the heavy nuclei, like many minority groups, contain a disproportionate amount of the information about their origin.

The cosmic ray energy spectrum, expressed as the flux of particles above a given energy E (measured in GeV) is approximately $(1+E)^{-1.5}$ up to about $10^{19}$ eV where the exponent approaches $-2.3$. At $10^{19}$ eV the flux is roughly $J=5 \times 10^{-18}$ particles $cm^{-2}$ $sec^{-1}$ $ster^{-1}$; thus, to count one of these particles per day one must place an array at the top of the atmosphere that is approximately 30 km on a side.

GALACTIC COSMIC RAYS

If the energy spectrum is integrated over all energies greater than 1 GeV, the result is an energy-density of $W_{CR}=0.9$ eV cm$^{-3}$, which is a rather modest figure. If the galaxy occupies a volume of approximately $10^{68}$ cm$^3$ the total energy W of cosmic rays contained within our galaxy would then be of the order of $10^{56}$ ergs—a very large number. Beyond the orbit of the earth the measured flux and energy spectrum of cosmic rays do yield an energy density of approximately 1 eV cm$^{-3}$, and it seems perfectly reasonable to assume that this energy density is spread throughout the galaxy. Thus, one must find a source that supplies a total cosmic ray energy of $10^{56}$ ergs and continues supplying it at a rate of the order of $10^{42}$ ergs per second. Now we should note that, for example, the energy density of starlight is on the order of 0.3 eV cm$^{-3}$. The x-ray and gamma ray energy is quite small, about $3\times10^{-5}$ eV cm$^{-3}$. The $3°$K black-body radiation has 0.4 eV cm$^{-3}$. Radio waves and neutrinos together have about 0.9 eV cm$^{-3}$, and the galactic magnetic field has approximately 1 eV cm$^{-3}$. The kinetic energy of the gas clouds is approximately 0.3 eV cm$^{-3}$. This suggests that either we have a series of coincidences, or we have some type of equipartition. It is difficult to see physically how equipartition can exist; thus, we have simply a perspective on what magnitude of energy density we are dealing with throughout our galaxy. We see that the cosmic rays take on major astrophysical significance through their total energy.

This summary can be concluded by giving a very broad phenomenological view of what we think is the life history of a typical series of cosmic rays. We start with acceleration in a region whose location we do not know—a "black box" to give the energy spectrum and the flux as they are observed. These particles will now diffuse through our galaxy and by this process the nature of the beam is changed. It can be assumed that while the flux is low and the density of interstellar material is quite small (1 atom cm$^{-3}$), the distances are correspondingly great, thus, some of the charged nuclear particles will undergo nuclear interactions, and the heavier particles will undergo fragmentation. This will introduce into the beam elements that are assumed not to be there initially, such as lithium, beryllium, boron, helium-3, and deuterium. These are assumed to be absent or very scarce in the source region, from what is known about universal abundances. One would expect that as carbon, nitrogen, and oxygen traverse the interstellar material, fragmentation products would be introduced. Interactions with the electrons also should introduce energy losses which modify the beam.

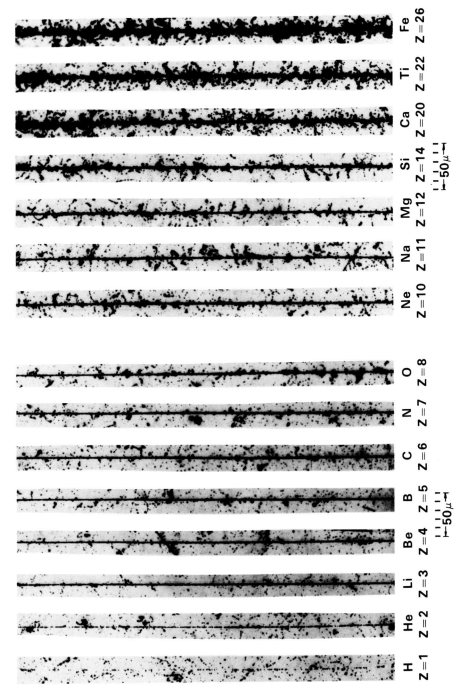

Fig. 7-1. Tracks of typical stripped nuclei as recorded in nuclear emulsions. (After Powell, Fowler and Perkins.)

When these cosmic rays approach the solar system they undergo another process at distances of 40 or 50 A.U. in which they diffuse inward against the solar magnetic field carried outward by the solar wind. If there were simply a solar wind expanding outward from the corona with no magnetic field, there would be no effect on the charged particles. Again, if the magnetic field were perfectly smooth, it would have no effect. But the magnetic field of the surface of the sun is irregular; thus an incoming particle sees a series of jagged magnetic field regions moving outwards. In its frame of reference it sees these as scattering centers which tend to sweep out the low-energy cosmic rays thus accounting for solar modulation. Thus at present one can only infer a value for the interstellar density which is believed to be valid; but we would like very much to get out to large distances of the order of 10 to 28 A.U. to make direct measurements.

HEAVY NUCLEI

It is appropriate that this is the subject matter for the conference because the galactic cosmic rays give a sample of matter from the far reaches of our own galaxy. Such matter has traversed most of the galaxy many times and has a significant amount of information if one knows how to decode it. The sample arrives in the form of fully stripped nuclei. One can define a minimum energy of approximately 10 MeV per nucleon which corresponds to particles of a minimum velocity of approximately 0.25 c. On the other hand, more than half will have energy greater than 1 GeV and therefore have a velocity greater than 0.9 c. Our sample arrives at very high velocity.

To define the chemical constituents in the sample one must deal with individual nuclei. What happens as these multiply-charged nuclei traverse matter —whether it is the matter in our detectors or the interstellar material or the material of the upper atmosphere? There are two primary interactions. The first is simply the coulomb interaction with the electrons in the stopping material; this and the coulomb interaction will increase with higher Z, so that one will observe more and more high energy electrons ejected by the passage of this multiply charged particle through material. This is the predominant interaction. The other interaction is the nuclear process in which the incoming primary nucleus will fragment and break up into constituents which continue to travel with essentialy the same velocity as the parent particle.

An example of quantitative analysis on individual nuclei is shown in Figure 7-1. These are tracks of cosmic ray primaries in Ilford G-5 emulsions as published by Powell, Fowler, and Perkins. They all have velocities such that

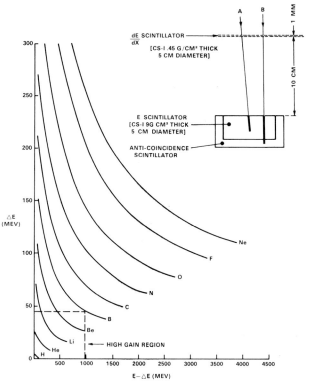

**Fig. 7-2. Calibration curves for energy and charge separation produced by the basic mass spectrometer shown in the upper right corner.**

$\beta=1$. There are examples of tracks of particles having $Z=1$ through $Z=12$, followed by $Z=14$, 20, 22, and 26. The proton is barely distinguishable at minimum ionization. As Z increases, so does the ionization. The iron nucleus has a track of hairy-like appearance produced by the coulomb interaction with the electrons. It transverses the emulsion undergoing collisions with electrons and ejecting them from the path, and these appear as delta rays or knock-on electrons; in fact the traditional way to determine the charge of the primary is simply to count the number of knock-on electrons above a certain energy along the track.

One very simple cosmic ray detector used in satellites by ourselves and by the University of Chicago is shown in the upper right-hand corner of Figure 7-2. One uses a thin counter to measure the rate of energy loss dE/dx and then looks at these particles which stop in the lower detector. One verifies that they did stop by surrounding this with an anti-coincidence counter. Measuring the rate of energy loss in the thin counter and the total energy loss in the lower detector constitutes, effectively, a mass spectrometer. The rate of energy loss is proportional to $Z^2/\beta^2$, the total energy is proportional to $m\beta^2/2$. Their product is a fraction which is proportional to $Z^2m$, so one expects to do both the energy and charge separation as shown by the family of curves in Figure 7-2 for the particles which stop. We place this device in a satellite outside the earth's magnetic field and then simply study those particles which stop in the detector, that is, particles from about 15 MeV per nucleon up to about 100 MeV per nucleon.

The raw data show the resolution obtained as illustrated in Figure 7-3. He$^4$ and He$^3$ are resolved quite cleanly. At the outset we expected no He$^3$; we now

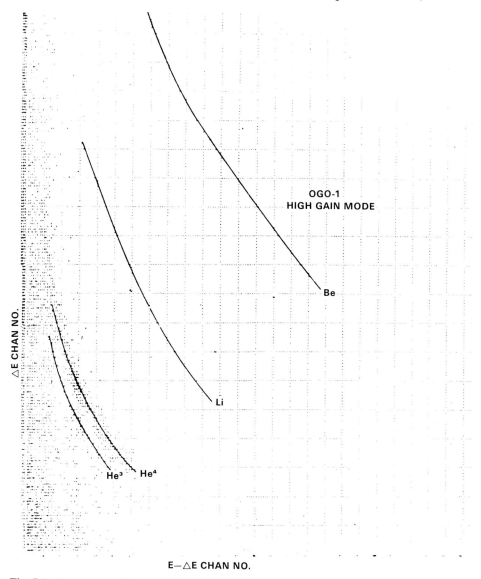

Fig. 7-3. An example of the raw data obtained by mass spectrometer in space.

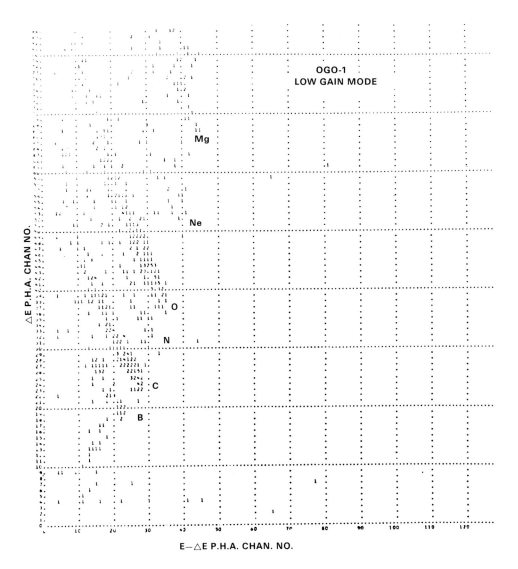

**Fig. 7-4. Charge spectrum analysis of primary cosmic rays. Lines corresponding to C, O, Ne, and Mg are clearly visible.**

believe it is introduced by the interactions of heavier nuclei in passing through the interstellar material. Figure 7-4 shows the lines corresponding to carbon, oxygen, neon, and magnesium. Chemical analysis of individual nuclei is easy to do if one can stop them.

Another technique, used at higher velocities, is based upon the Cerenkov counter. Particles with velocities greater than the velocity of light in a medium produce an output from the counter that is proportional to $Z^2(1 - \frac{1}{\beta^2 n^2})$

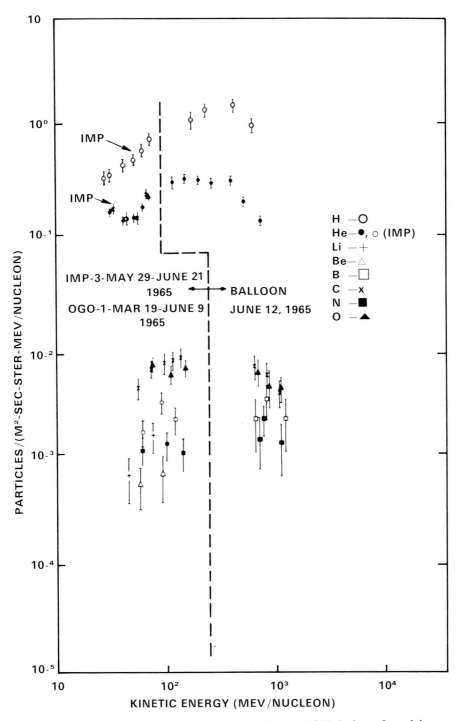

**Fig. 7-5. Typical energy spectrum from satellite flights in 1965 during solar minimum.**

where Z is the charge of incoming particle which passes through material of index refracton n. This particular combination can isolate the relativistic nuclei from those of lower energy. For example, lucite has n=1.5, which means that for $\beta \leqslant 0.66$ no light emerges from the Cerenkov counter. This

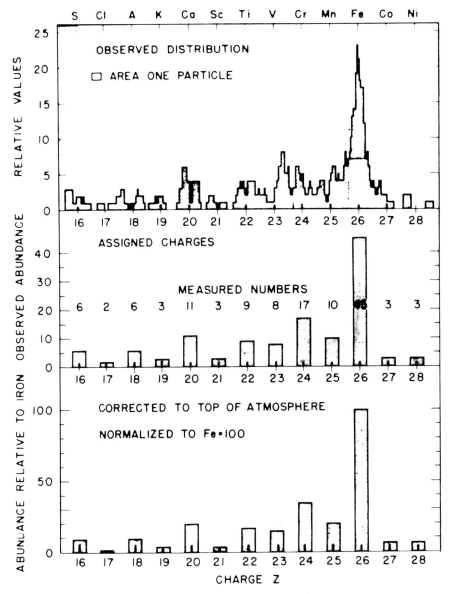

**Fig. 7-6. Raw data on cosmic ray primaries in the region of iron as recorded in nuclear emulsions. (After University of Wisconsin.)**

provides a means of charge analysis and energy analysis of the high-energy particles. A typical energy spectrum from the satellite flights shown in Figure 7-5 was obtained during minimum solar activity in 1965.

The flux of particles is remarkably small between 30 MeV and 1 GeV. A more significant parameter is the relative abundance of these elements. Figure 7-6 supplied by the University of Minnesota group shows the raw data in the region of iron; we see hints of individual resolution in the region up to iron. For many years people looked in vain for cosmic rays of Z greater than iron until we began to call it the cosmic ray iron curtain, but this curtain has now been penetrated.

An example of a relative abundance study by the University of Chicago is given in Figure 7-7. The abundances are normalized to carbon and oxygen. There are three remarkable features. First, there are surprising abundances of Li, Be, and B, comparable in fact to the abundances observed for Ne, Mg, and Si; and these latter reasonably approximate the universal abundances.

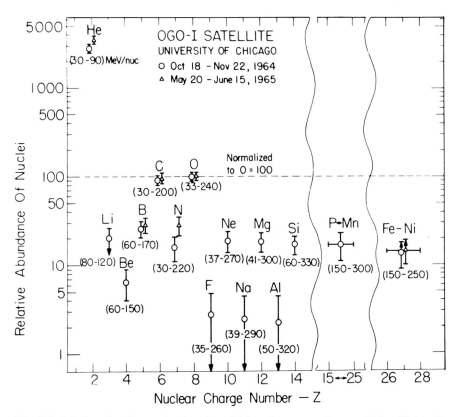

Fig. 7-7. Relative abundance of stripped nuclei in primary cosmic rays. (After University of Chicago.)

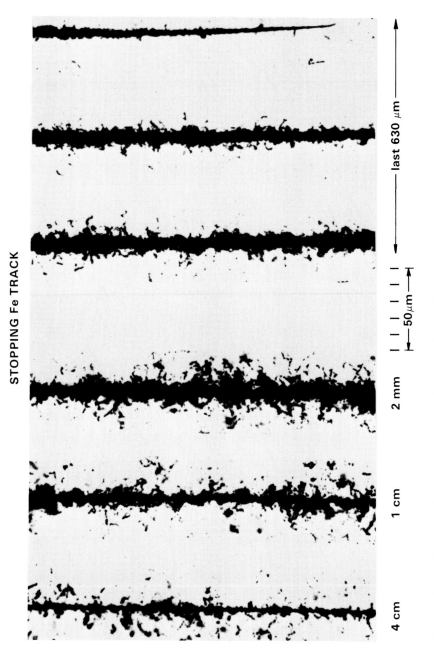

Fig. 7-8. A stopping Fe nucleus showing typical thin-down due to electron capture-and-loss. (After University of Bristol.)

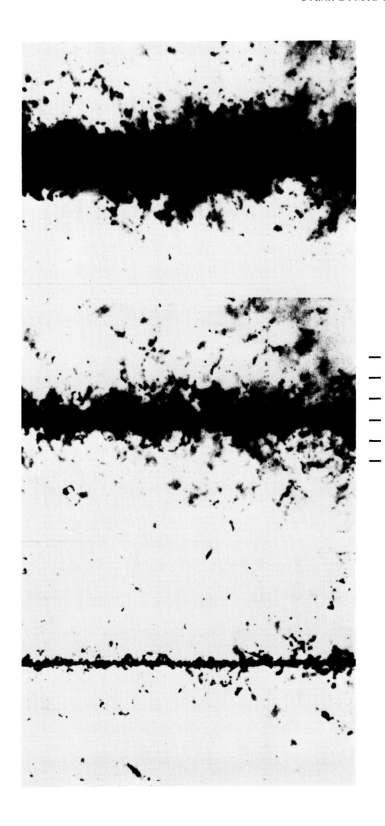

50 μm

Fig. 7-9. Track of nucleus of Z~90 in nuclear emulsion. (University of Bristol.)

These light elements are supposed to be three orders of magnitude below the medium group in abundance, yet they are comparable. The most plausible explanation is the process of fragmentation through the nuclear interactions of the original cosmic ray beam with the interstellar hydrogen. If one knows the probability of such fragmenting, one can deduce the amount of material traversed by the Li, Be, and B. The second feature is a very strong abundance of iron. Third, in this region where there is good charge resolution there is a very pronounced even-odd effect: the abundance of the even nuclei is greater than that of the odd. Elements produced by fragmentation should not exhibit this effect, and this argues that most of the Li, Be, and B arises from the material at the source. Recent new information from solar modulation provides means to obtain a preliminary calibration of the source and perhaps resolve this conflict. More on that later.

ULTRAHEAVY NUCLEI

One of the more exciting recent discoveries is the work of the Bristol group in extending the charge spectrum beyond iron. As a normalization point we examine Figure 7-8 which shows their picture of an iron nucleus in its last 4 cm before stopping. It produces maximum ionization about 2 mm before it stops, then it starts to "thin down," and finally comes to rest in the emulsion. During the "thin-down" it slows sufficiently to acquire some orbital electrons; it may lose a few of them, but on the average its net charge decreases gradually and the thickness of the track decreases until the multiply charged nucleus stops.

The Bristol group imposed some 3 or 4 $gm/cm^2$ between two slices of emulsion to obtain their famous photograph. Figure 7-9 shows what happens in 1 cm. The iron nucleus is obvious, but in addition there is a pronounced "tunnel" through the emulsion, produced by a nucleus with Z in the range from lead to uranium. So the English have found a new source of uranium that has a flux of the order of $10^{-8}$ particles $cm^{-2}$ $sec^{-1}$ $ster^{-1}$. The exact charge identification is uncertain, but they established that two of these nuclei, including this one, came in and did not slow down in traversing 4 or 5 $gm/cm^2$ of material. For reference, the relativistic iron nucleus is $Z=26$ and their best analysis established this as a charge in the range of 80 to 90.

How did they do this when other people had not seen it before? Instead of flying small emulsion stacks of 6 or 8 in. on a side, they flew 5 sq m of emulsion. This is an enormous scanning task, but tracks as heavy as this can be found with very small magnification. Bristol now has twelve of these tracks

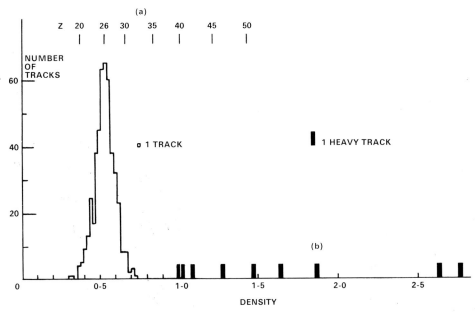

**Fig. 7-10. Histogram of "very heavy" and "very, very heavy" nuclei showing twelve tracks between Z=82 and Z=92. (After University of Bristol.)**

in the lead–uranium region as shown in Figure 7-10. Others appear in the Z>36 region, and of course the iron peak is at Z=26.

One other previous report of nuclei in this range is noteworthy and is one of the interesting new developments from the study of meteorites. Price, Fleischer, Walker, and others at General Electric have shown that if a heavy primary charged particle passes through an insulating substance it leaves a trail of damage to the crystalline lattice structure that is not easily erased over time intervals of the order of $10^6$ to $10^7$ years. The insulating substance is cleaved and exposed to an acid bath, which etches preferentially along the damage left by the heavy primary. Other sources of heavy ionization such as spontaneous fission will leave permanent tracks. These are called "fossil-charged particle tracks." An example of the tracks found in meteorites is shown in Figure 7-11. Two methods of identification have demonstrated that they are charged particles entering from outside the meteor; one, the very rapid attenuation observed from outside of the meteor inward—the aniso-tropic angular distribution; and two, the fact that the tracks clearly look different—some are appreciably longer. The GE group was able to estimate a flux for Z>32 by machine calibration. Another example of this technique in an olivine crystal is shown in Figure 7-12. One again sees these long etched tracks in a sextacleavage plane. This is the least expensive particle detector acting over the longest period of time yet devised. Their results for the flux of elements above $Z=32$ are not in strong disagreement with what has been found for heavier elements. Walker *et al.* have named the iron group "very heavy" and the $Z>32$ "very, very heavy." (I rather suspect that had the British established priority we'd probably be talking about the rather heavy and the quite heavy, but the course of empire was reversed.)

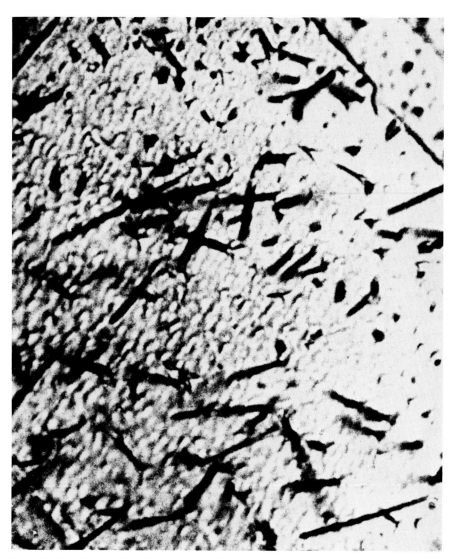

**Fig. 7-11. Fossil-charged particle tracks of large Z observed in meteorites. (After Walker et al.)**

**Fig. 7-12. Tracks of heavy particles in olivine crystal. (After Price et al.)**

SOLAR COSMIC RAYS

Ideally we would like to have a calibrated source, but we do not know the origin of the galactic cosmic rays. On the other hand, we have a near neighbor, a large star, the sun, which obliges us from time to time with large solar flares. Figure 7-13 is a montage of pictures by Fichtel *et al.* where it is seen to be far simpler to get a direct sample off the solar surface than it is to get a lunar sample back. Fichtel and others fire rockets a few hours after a large flare on the sun. The first part shows the situation when there are no solar particles. Next is a corresponding five-minute rocket launch and exposure through the November 12 event. Finally the figure shows the very steep energy spectrum and the fact that these tracks can be observed in a relatively easy way.

The most striking feature was the family of heavily ionizing particles. Figure 7-14 shows a calcium nucleus coming to rest among the background tracks. This was typical of the rocket emulsion samples; they would see none, or maybe one or two of these heavy stopping tracks. Better results were obtained when the rockets were fired after the monitor—the satellites, the riometers, and the other indicators—reported a large flux of energetic particles put out by a flare. Thus the elements of this solar surface sample can be identified by the nuclear emulsion techniques. The most accurate determination of the amount of neon relative to oxygen and some of the heavier elements were

**Table 14. Solar Cosmic Ray Abundance Normalized to Oxygen Compared with Photospheric Abundances Not Normalized to Oxygen.**

| Element | Solar Cosmic Rays | Solar Photosphere |
|---|---|---|
| $^2$He | $107 \pm 12$ | |
| $^3$Li | | $< 10^{-5}$ |
| $^4$Be-$^5$B | $< 0.02$ | $< 10^{-5}$ |
| $^6$C | $0.59 \pm .07$ | $0.60 \pm 0.10$ |
| $^7$N | $0.19 \begin{smallmatrix} + .04 \\ - .07 \end{smallmatrix}$ | $0.15 \pm .05$ |
| $^8$O | 1.0 | 1.0 |
| $^9$F | $< 0.03$ | 0.001 |
| $^{10}$Ne | $0.13 \pm .02$ | 0.11 |
| $^{12}$Mg | $.042 \pm .011$ | $0.051 \pm 0.015$ |
| $^{14}$Si-$^{21}$Sc | $.090 \pm .020$ | $0.097 \pm .003$ |
| $^{22}$Ti-$^{28}$Ni | $< 0.02$ | 0.006 |

made by this method. Now the spectroscopic values have slowly started to edge into agreement with the results obtained from nuclear emulsions.

The Goddard Space Flight Center emulsion group has found the same solar composition for five such exposures. It gives them faith that they are seeing a representative sample of the solar surface accelerated by the flare. There is good evidence at the moment that the particles are accelerated to a few MeV per nucleon over a period of hours or days or weeks before the flare; they are trapped in radiation belts above the active region and the action of the flare is to raise their energy by a factor of 10, at the same time sweeping them out from the sun. In any event, it is a reproducible phenomenon. Table 14 shows the solar cosmic ray abundance normalized to oxygen compared with the photospheric abundance not normalized to oxygen. There is very good agreement in the elements of nitrogen, fluorine and neon (which were not known too accurately), and quite good agreement for magnesium and the heavier elements. One would expect to improve this technique and extend it to higher energies and higher charges.

It has been established that the ratio of lithium, beryllium, and boron to carbon, nitrogen, and oxygen is about 0.3 in the galactic cosmic rays. From the argument about the solar cosmic rays, I believe that the carbon, nitrogen and oxygen are representative of the source itself. Since the flux of iron, and of carbon, nitrogen, and oxygen are known, one can calculate through how much hydrogen the carbon, nitrogen, oxygen, and heavier elements must be present to produce the observed lithium, beryllium and boron. The result is a path length of about $4 \pm 1$ gm cm$^{-2}$. The figure was formerly $2.5 \pm 0.2$; it has risen in the last year or so because the nuclear physicists have given much better fragmentation parameters. The probability that carbon will produce

SOLAR COSMIC RAY PRODUCING
FLARE ON SUN

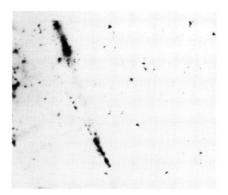

NORMAL COSMIC RAY EMULSION
EXPOSURE

SOLAR BEAM EXPERIMENT
LAUNCHING

NOVEMBER 12, 1960 EXPOSURE
1 mm FROM OUTER EDGE

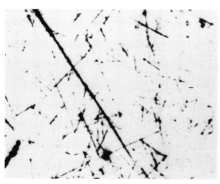

SOLAR COSMIC RAY HEAVY NUCLEI

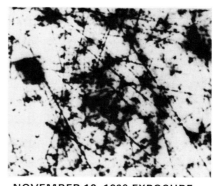

NOVEMBER 12, 1960 EXPOSURE
20 mm FROM OUTER EDGE

**Fig. 7-13. Montage of nuclear emulsions flown in rockets to record cosmic ray tracks, some of which are from the sun. (After Fichtel.)**

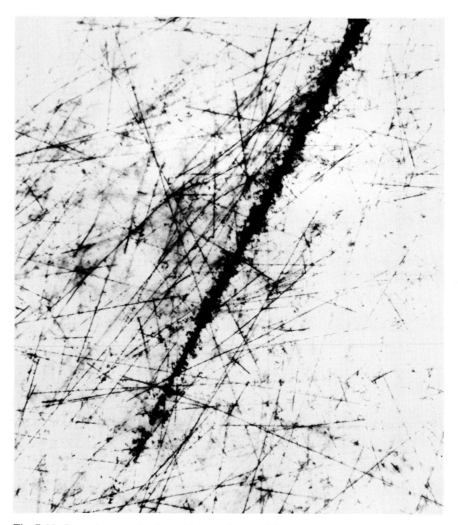

**Fig. 7-14. Ca nucleus coming to rest in a rocket emulsion sample.**

lithium in nuclear interaction has become much better known, primarily because of the work of the group at Saclay. The 4 gm cm$^{-2}$ will produce the flux of helium-3 and deuterium that one observes. From the average path length of travel, the average density, assuming the velocity is a c and using the 4 gm cm$^{-2}$ the mean lifetime is on the order of $2 \times 10^6$ years for a density of 1 atom cm$^{-3}$. The result is strongly dependent upon the average density, which is not known very well—whether it is 1 atom or 0.01.

What is the origin of galactic cosmic rays? We simply do not know. Many investigators have looked at the most active objects in the sky, such as supernovae and the like, as proposed sources. Everywhere one turns, nature is able to accelerate particles with really surprising ease—the galactic cosmic rays, the solar cosmic rays, the earth's radiation belts. We have our own private accelerating mechanisms here on earth and yet we do not understand a single one. The same thing is true concerning the galactic cosmic rays.

There are two contending schools of thought at the moment. One is the Russian school led by Ginzburg who says that the galactic cosmic rays must be confined within our own galaxy. To account for a $10^{56}$ eV total energy requires a supernova to emit on the order of $10^{49}$ ergs every hundred years. The other school, that of Burbidge, Hoyle and others, says that for the path length calculated a fast cosmic ray nucleus would travel a hundred times from one side of the galaxy to the other in its mean life; thus there is no reason to expect that the galaxy, with its not very effective trapping mechanism, can confine this cosmic ray, and the metagalactic space is filled with cosmic rays. The source of the cosmic rays then becomes the large radio galaxies that one sees in profusion. Clearly it is not known whether the galactic picture or the metagalactic picture is correct. Whether they are galactic or metagalactic material, cosmic rays offer the only direct sample of extra-solar system.

DISCUSSION

(Q) Are all the tracks in Figure 7-1 those of incoming particles or are some of them produced in the emulsion?
(A) McDonald: This is the incoming particle in each case. It is simply the track as recorded in the emulsion. It gives graphic evidence of how the track of a particle appears to any type of ionization detector. We would expect that these particles, at energies above a few GeV, would have an overwhelming probability to undergo nuclear interactions before they stop. At lower energies—a few hundred MeV and below—it is much more probable that they will simply stop.
(Q) Referring to the graph of energy loss vs energy, there was a cut-off on the high energy side. Is this because the ranges were greater?
(A) McDonald: No. As you go to higher charge two things occur: the threshold is lower and the energy range broadens out. Only those particles which stop in the detector are analyzed. It depends on the charge. All these curves were drawn for one thickness of detector.
(Q) Did the Bristol group find a lower flux than the Price, Walker, and Fleischer group for the very, very heavy or the rather heavy?
(A) McDonald: The Bristol group found a higher flux, but Walker *et al.*

were very cautious, and I think properly so, although I think it is a beautiful piece of work. They really lack absolute calibrations in the higher Z range. (Q) What is the status of limits on anti-particles in cosmic radiation? Would you care possibly to comment on whether or not this might be related to the last question which you were discussing concerning the sources of the primary cosmic rays?

(A) McDonald: Looking for anti-particles is a very worthwhile endeavor. In cosmic rays, there seems to be a chance to sample material that probably or possibly comes from outside our own galaxy, and since there is no way to prove that the other galaxies are not made up of anti-matter, it is definitely worthwhile looking for it. As far as a source is concerned, one can produce 10-GeV particles from the sun, although the sun is not necessarily a typical example. We would expect flare stars that are many, many orders of magnitude more active which might be suitable sources. It is not necessary to involve anti-matter in the acceleration. Looking for anti-matter is a perfectly proper pursuit in its own right, which one ought to do, and it is probably not very difficult if there is any reasonable abundance like a fraction of a percent.

(Q) What is the latest evidence on what happens to the curve of dE/dx as a function of E at large relativistic energies? There has been some uncertainty whether that curve flattens out or continues to rise weakly.

(A) McDonald: Yes, I think for the most part that is not a problem here because one doesn't know the charge spectrum, with any great precision, of particles with energies greater than say, 10, so you're not in a region where this is important. Actually, one of our main programs at the moment is to look at the charge spectrum for about 100 GeV up to something on the order of $10^4$ GeV. We want to look at the Cerenkov light and look at the dE/dx in thick ionization detectors, where one suspects that the relativistic increase is only a small effect of about 6 to 8 per cent. Then, with an ionization calorimeter where one stacks up, say, 4 or 5 tons of iron interlaced with various materials and scintillators, we want to run a three-dimensional air shower experiment by looking at the cascade of ionization through this as it builds up and dies out. In this way we can measure the total energy by measuring the total ionization. The Cerenkov light should be linear. There have been recent experiments with xenon looking at very fast mesons where a very definite rise is detected—the dE/dx with E going up about as you would expect—so the relativistic rise is there in gases. It is probably a small 6 or 7 per cent effect if you stay with solids. If we extrapolate our energy spectrum, we always worry about this and it doesn't look like a large effect until one gets up to appreciably higher energies. This would contribute something to understanding the density effect at these higher energies.

(Q) What is the present status of favor or disfavor concerning Fermi's accelerating mechanism for cosmic rays?

(A) McDonald: I don't think there is any. The general feeling is that we have a number of situations where nature is accelerating particles: our own magnetosphere, the sun, the galactic cosmic ray sources. My view is we don't understand any of them. We don't know the mechanism by which these particles get accelerated. Some experimenters have concentrated on the magnetosphere while we have concentrated on the sun. The basic conditions in terms of the magnetic fields simply are not known—how they change and what are their time derivatives. The whole acceleration question is completely up in the air at the moment. It may turn out that the Fermi theory is valid for some special configuration. The idea that I personally favor at the moment for the sun is a two- or three-stage process not unlike the operation of large machines where you inject particles initially at a few MeV. Why should nature be different? I feel that there is a very strong evidence to support this. The particles simply cook for a period of a few days or a few weeks, building up slowly from thermal energy to 3 to 30 MeV. Finally the flare comes along and increases their energy by merely a factor 10. So instead of taking a particle from thermal energy up to 300 MeV in one process, we take it from 30 to 300 MeV, which makes a much easier problem to solve.

(Q) What is the latest evidence on the existence of charges of ⅓ and ⅔ electron charges in your experiments?

(A) McDonald: We have not looked for quarks but from talks I have heard at the University of Maryland and from Ted Bowen at the University of Arizona I have learned that the upper limits on the quark population are getting very, very small. Our next cosmic ray experiment has a size of about a half meter by a half meter, whereas people already looking for quarks have apparatus 3 or 4 m on a side, and so they have more problems than we have.

# IV. Cosmic Rays and Atmospheric Neutrons

# The Electron Component of Cosmic Rays PETER MEYER

*Abstract: Only in the past six years has the presence of electrons in the cosmic radiation been firmly established. Measurements at balloon altitudes made with electron spectrometers of new design have shown that the electron component is relatively insensitive to solar modulation, has less positive excess than simple theories predict, and exhibits a change in the power law spectrum at high energies. Further studies of cosmic ray electrons may reveal the characteristics of the source of cosmic rays and the details of their storage mechanism in the galaxy.*

INTRODUCTION

There has already been a discussion of the various cosmic ray nuclei which we believe to originate in the galaxy and arrive at the earth, particularly the large abundance of protons, somewhat smaller abundance of alpha particles, and the small abundance of heavy nuclei consisting approximately of 1 per cent of the total proton flux. In spite of the small intensity of the heavy nuclei, they are studied very intensively and, as Dr. McDonald pointed out, since they contain information which one could not decipher if one would look only at the protons. They reveal much about the chemical composition of the sources of cosmic rays and, equally interesting, about the interactions which the nuclei undergo on their way between their sources and their arrival in the vicinity of the earth.

Electrons and positrons are as rare as the heavy nuclei—again they amount to roughly about 1 per cent of the flux of the protons. Nevertheless, they have been studied very intensively in the last years. The reason is exactly the same as the reason for the interest in the heavy nuclei, because this component has a number of unique features which promise to reveal information concerning the origin and the storage of the cosmic rays in our galaxy. It is these two problems to which I would like to address myself and I shall give you a summary of what we have learned and what our thoughts are on the basis of the experiments on the electron component.

SOURCES

The regions in which there may be cosmic ray sources are divided into three parts: one is well known—the sun; the second may be our galaxy; and the third all the rest of extra-galactic space. There is good evidence today that all three regions are sources of energetic particles. The sun can be eliminated from discussion immediately because, although it is a producer of nuclei and electrons, we know now that it is not a major contributor to the over-all level of cosmic radiation, particularly at high energies. This leaves the galaxy and the extra-galactic space. It is also quite certain today that both these regions contribute to the particles seen here within the solar system. The work that was done in the past years on extremely high-energy cosmic ray particles in the range from $10^{17}$ to $10^{20}$ or $10^{21}$ eV, and the discovery of the near-isotropy or possible complete isotropy with which these particles arrive at the earth, have made it clear that they cannot possibly be stored within galactic magnetic fields and therefore must be of extra-galactic origin. But this is a small fraction of the total cosmic radiation.

The bulk of the radiation has much lower energy, and most of us believe today that this bulk of the cosmic radiation has its source within our galaxy. Observations of the nuclear components cannot very easily confirm this belief. However, if one looks at the electrons, one could come to quite different conclusions. Probably all of them are accelerated within the galaxy. The main clue to this conclusion comes from the charge distribution of the electron component. The fact that electrons are present in the galaxy was first noticed by radio astronomers through the observation of a widely distributed frequency spectrum which could not be reconciled with a thermal origin but which must be due to some nonthermal processes. This source is distributed over the galaxy and and possibly a region surrounding the galaxy. The radio astronomers in the early fifties made measurements of the radio intensity from this region and found that signals came from an almost spherical volume around the galactic disc which was given the name of galactic halo. The interpretation of this radiation came very soon after its discovery; the radiation was interpreted as being due to relativistic electrons which move in magnetic fields on spiral paths and which therefore emit synchrotron radiation. The radio spectrum that was observed at the earth is very compatible with this type of a conclusion. It was then immediately asked where do the electrons come from, and the obvious suggestion is that, since there are cosmic ray protons in the galaxy and there is also interstellar hydrogen gas, high-energy collisions between protons and protons must lead to the reaction producing two nucleons plus multiple pions.

$$p + p \rightarrow 2N + n\pi .$$  Equation 1

The $\pi$-mesons will decay therefore by the familiar process $\mu^\pm \rightarrow e^\pm + \nu + \nu$. As a consequence one should see secondary galactic electrons and positrons.

The energy spectrum of the protons is well known. There is also a good estimate of the density of matter within the galactic disc. Furthermore, from accelerator experiments, the cross-section for the production of these pions is known and therefore a good estimate can be made of the source intensity for the electrons and positrons. One can also make an estimate of the fraction of positrons to electrons. The source spectrum log I vs log E would have a broad peak around 100 MeV and then goes over into a power law spectrum with a shape equal to the shape of the protons which produce the electrons. Calculating the positron fraction from the collision process yields

$$f^+ = \frac{N^+}{N^+ + N^-} = 0.65$$  Equation 2

in an energy range of 1 to 10 BeV. I chose this energy range because this is the one that can be compared with experiments. At much higher energies this number will tend to go toward 0.5. The reason for a positive excess is simply that one starts out here with two positively charged particles. In 1964 Mr. DeShong at Argonne and Dr. Hildebrand and I were able to show experimentally that this fraction $f^+$ in the above energy region (it was a more limited energy region at that time) was certainly appreciably smaller than 0.5.

In the past year this experiment has been very much refined, mainly by a student of mine, Dr. Hartman, who obtained quantitatively the fraction of the positrons in the primary electron flux as a function of energy. Figure 8-1 shows the apparatus that is used to make such an experiment. This is the

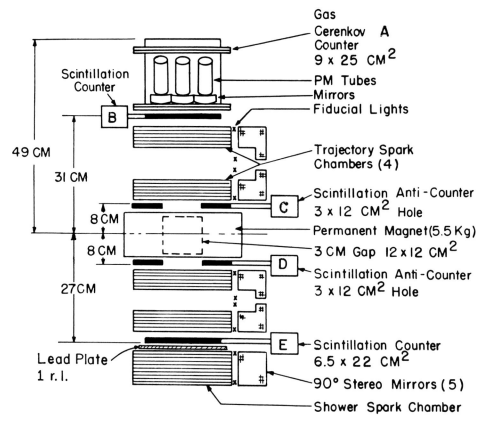

**Fig. 8-1. Electron spectrometer flown to balloon altitudes at Fort Churchill. Charge separation occurs in the magnetic field and the electrons produce characteristic showers in the heavy plate chamber. Selection is made by counter telescope, Cerenkov counter, anti-coincidence counter and spark chambers.**

equipment which is flown to balloon altitude at northern latitudes of Fort Churchill. To separate the charges a magnetic field is needed, and the heart of the experiment therefore is a permanent magnet through which the particles move. They are selected by a counter telescope, gas Cerenkov counter, two scintillation counters, and two anti-coincidence counters which have a hole just large enough to let the particles pass through the magnetic gap; no counters are put into the gap because one wants no scattering there. Five spark chambers are arranged in this equipment so that we can see the trajectory of the incoming particle coming through the field. The magnetic deflection is given by the charge as well as the rigidity or energy, and there is a heavy plate chamber in which the electrons make a characteristic shower. The spark chambers are looked at from the front and also from the side.

Figure 8-2 shows a picture of a 400 MeV electron taken at altitude during

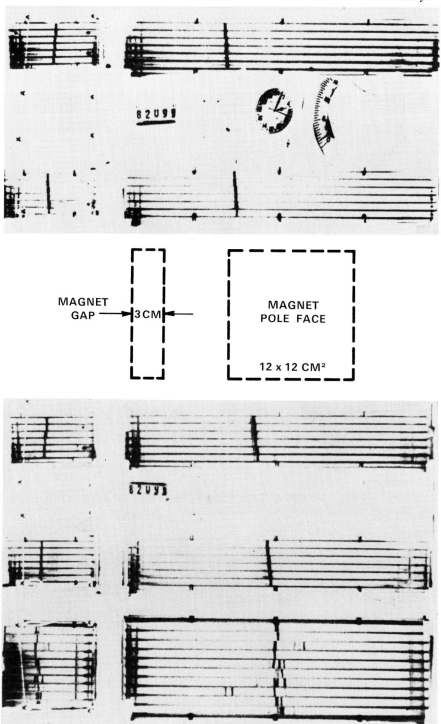

**Fig. 8-2. A 400-MeV electron event at balloon altitude.**

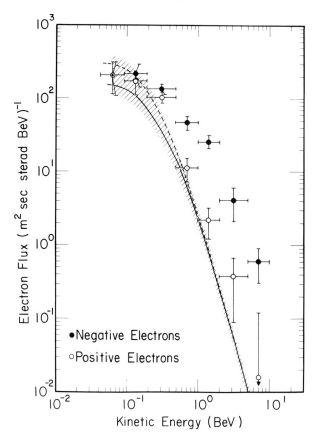

**Fig. 8-3. Electron flux as a function of energy. (After Hartmann.)**

the flight. One can see with the naked eye that there is a deflection. At about 500 MeV the fraction $f^+$ is of the order of 10 per cent or less—very far from 0.65. Predominantly negative electrons constitute the primary electron components. The most recent experiments give a value of 0.06 between 5 and 10 BeV. This agrees quite well with measurements which were made by a group in Italy and France in a joint experiment. It disagrees to some extent with measurements done by a group at the Tata Institute, although their work was at higher energy. We conclude that the electron component consists mainly of negative electrons and that the collision process in Equation 1 plays only a minor role in the contribution to electrons. Figure 8-3 is the energy spectrum which Hartman has deduced from his measurement. The negative electrons are the dark dots and the remainder are the positive electrons. Measurements below approximately 500 MeV are ambiguous because of the background of atmospheric secondaries which even at balloon altitudes produce many more electrons than there are in the primaries.

The curve is the theoretical calculation which would give the positron spectrum produced in the galaxy. These positrons certainly have to be there because it is known that they are cosmic rays, that there is the interstellar gas, and within the large errors there is no disagreement that indeed this number of positrons is present.

Since the proton collisions appear inadequate to account for the source of electrons, it would be well to examine a suggestion made more than ten years ago by Ginzburg and by Hayakawa that type 2 supernovae might be responsible for the emission of the cosmic radiation. Optical and radio-astronomical observation of supernova remnants have shown clearly that high-energy electrons must be present in these objects. The most famous one is the Crab Nebula. In the nebulosity regions light is emitted with a continuous spectrum and this light is polarized. In view of this evidence it is absolutely clear that there is a source of synchrotron radiation, which means that high-energy electrons are present in such objects. The energy has to be at least $10^{11}$ eV, and if some of the x-rays observed are also due to synchrotron radiation the energy must be two orders of magnitude higher. This means that the lifetime of the electrons is short compared with the lifetime of the supernova remnant and that they have to be accelerated and reproduced constantly. These electrons probably originate in the tenuous plasma which forms the supernova shell. It is likely, therefore, that they are negatively charged, which would be in agreement with the observations. So, as far as the electron component is concerned, one has come close really to a verification of the suggestion of the supernova origin.

Why must one rule out the possibility of an extra-galactic origin of the cosmic ray electron component? Again, one deals with two possible sources. One is collision between high-energy particles and the extremely thin gas in extra-galactic space. If it were collisions, again one would get a positive excess so one can rule this out immediately. The second reason is the presence of the universal black-body radiation corresponding to approximately $3\,^{\circ}$K. which Dicke has postulated to be the remnant of a primordial fireball originally at very high temperature in equilibrium with the matter, decoupled from the matter, and then cooled adiabatically to $3\,^{\circ}$K. because of the expansion of the universe. The evidence is very strong that this radiation exists, and if it does exist there is a large density of photons throughout space. The high-energy electrons would collide with these photons in the well-known Compton process. In the frame of the fast-moving electron, these photons look like gamma rays; therefore the collision cross-section is high and the electrons quickly lose their energy long before they have a chance to reach the vicinity of our galaxy from any other galaxy which could possibly be the source of such particles. We therefore have today in our hands strong experimental evidence for the galactic origin and acceleration of at least one of the cosmic ray components. Those who are inclined this way may take it as an indication of the galactic origin of the other cosmic ray components as well.

STORAGE

The question of the storage responsible for keeping the cosmic rays in the galaxy has long been debated by cosmic ray physicists. Today this question may be accessible to some new experimental tests through the investigations of the electron components. The background from which to examine this problem is the following. Through the study of Li, Be, B, $He^3$, and $He^4$ in the primary cosmic radiation it has been well established that the cosmic radiation has passed through approximately 3 gm $cm^{-2}$ of interstellar matter during its lifetime. Under the assumption that the density of the gas in the galaxy is of the order of 1 to 2 hydrogen atoms/cc, it can be concluded that these particles have lived approximately $10^6$ years in the galactic disc where the gas density exists. Of course, they may live for a much longer time in regions outside the disc where the matter density is very much smaller than this, possibly in a region like the halo or in a region where there are magnetic fields outside the galactic disc. In order to understand the isotropy of the very high-energy cosmic rays, whose radius of curvature in the galactic magnetic fields exceeds the dimensions of the galaxy, the radio-astronomers' halo has been invoked as an additional storage volume as proposed by Biermann and Davis. It has been assumed that the galactic halo contains magnetic fields of a few microgauss, but this is purely an assumption for which there is no direct evidence.

Although the question of the presence or the absence of such halo fields is very much under debate today, studies of the electron component can illuminate the question of cosmic ray storage. The reason is found in the difference of the loss processes which affect the electrons as well as the nucleons and nuclei. Both components eventually diffuse out of the vicinity of the galaxy— the diffusion process affects both components. For the nucleons and nuclei, this is about the only process by which they are lost. They undergo collisions in the galactic disc, but the total amount of matter traversed is about 3 gm $cm^{-2}$, and the collision mean free path is of the order of 50 gm $cm^{-2}$, so very, very few get lost. The electrons, however, have additional loss processes by collisions with photons from starlight as well as the black-body radiation and they move in the magnetic fields, thereby emitting synchrotron radiation. These loss processes are very powerful for electrons. The time rate of energy loss can be expressed by the equation

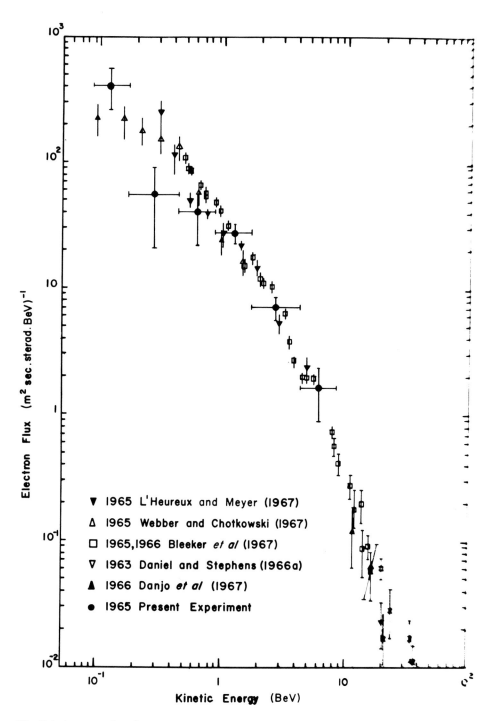

**Fig. 8-4. A composite of measurements of the energy spectrum of cosmic ray electrons from a few hundred MeV up to a few hundred GeV.**

$$-\, dE/dt \sim \left(\frac{B^2}{8\pi} + W_{ph}\right) E^2 = bE^2 \qquad \text{Equation 3}$$

where $B^2/8\pi$ is the energy density of the magnetic field and $W_{ph}$ is the energy density of the photons. It is clear that the high-energy particles are affected very much by these energy losses. Therefore, the source spectrum is modified due to such loss processes.

To find the spectrum in equilibrium at the point on the earth where one sees the particles in equilibrium with the loss processes, one has to solve something like a continuity equation. For the stationary state the time derivative of the number of particles must be zero. This can be summarized by the equation:

$$dN/dt = 0 = g(E,\vec{r}) + \frac{\partial}{\partial t}\left[\frac{dE}{dt}\, N\,(E,\vec{r})\right] + \nabla^2 DN(E,\vec{r}) \qquad \text{Equation 4}$$

which consists of a source term, an energy loss term and a diffusion term. This must be solved for N to give the intensity of the particles in the stationary state. It is a very miserable equation to solve, and it depends upon the model that one assumes. If one assumes a source spectrum of the form of the power law $E^{-\gamma}$ for the galactic disc, one would find that at a critical energy the spectrum gets steepened, and the equilibrium spectrum at the higher energies would have the form $E^{-(\gamma+1)}$.

What do the experiments tell about that? A composite of measurements of the energy spectrum of electrons by various groups in the energy range from a few hundred MeV up to a few hundred GeV is shown in Figure 8-4. It is quite amazing how well the spectrum is known today, considering that only six years ago primary electrons could be identified in the cosmic radiation. It is clear that this spectrum cannot be represented just by a simple power law over the entire energy region. Although I have drawn no lines through the data points, I could have convinced you that I could represent the spectrum quite nicely with two power laws—$E^{-1.6}$ at low energies reaching up to approximately 3 or 4 GeV and $E^{-2.6}$ at higher energies where the transition takes place at approximately 5 GeV. Does one obtain such a kink from the theory? Indeed one does; if we solve Equation 4 the first change in the energy does occur in a region from 2 to 4 GeV depending on what numbers one uses for the dimensions of the galactic disc.

At the present time, one is tempted to say that indeed such an interpretation is a reasonable one, but one should question what the implications would be. First, if this interpretation is correct, it would mean that the source spectrum of the electrons must be flat out to high energies. This is a highly unpleasant kind of conclusion because it is well known that the proton spectrum and the spectrum of the heavy nuclei has a spectral shape with an exponent of about

2.6. That means the electron source spectrum would be quite different from the source spectrum of the protons, and although not impossible, it is a conclusion one cannot accept very readily. That the electron spectrum does indeed extend with the same slope over such a wide energy range is a pure assumption at the moment. It could be that nature has just produced an energy spectrum which has a kink at this energy and that this interpretation therefore would be incorrect. A second objection or question can be raised. One knows very well that within the solar system where one measures these electrons,

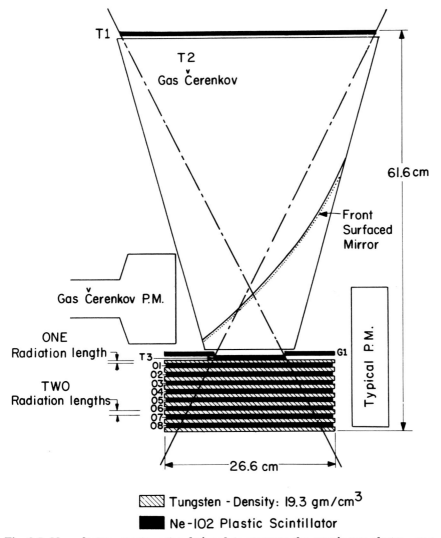

Fig. 8-5. New electron spectrometer designed to measure the cosmic ray electron component up to 200 GeV.

low-energy cosmic ray particles of all kinds are modulated by the sun. In particular, the low energies are suppressed in periods of high solar activity. How does one know that the change in spectral slope is not due to the modulation effect? During the past few years measurements of the spectral changes of the low-energy electron intensity have been looked for in order to sort out the modulation which is very predominant for protons and nuclei. Remarkably little modulation of the electron component has been found, and although this is a preliminary result, a group in Holland found a similiar result. One has the suspicion that there is relatively little modulation, particularly in the years of low solar activity when these measurements were taken. I am tempted to interpret these results in a preliminary way as telling the story that indeed there is some storage of the electrons outside of the galactic disc. I do not invoke a halo with an artificial boundary; it is perfectly sufficient to have a region of magnetic fields which is present ouside of the galactic disc. From considerations of the dynamics of the galaxy and the cosmic rays in the galaxy it has been postulated that at least there must be a fuzzy halo present around the galaxy.

Since I am an experimenter, I shall say a few words about further experiments to extend the spectrum to higher energies. It could be extremely interesting and important to see whether the spectrum shape persists or to see a steepening of the spectrum toward higher energies. This would yield more exact and detailed information concerning the points discussed before. Of course one would also like to extend the measurements to lower energies. At lower energies one has great difficulty in doing balloon work because of the secondaries in the atmosphere. Just two days ago the OEO-5 satellite was launched which carries our electron experiment on board and will help cover the region from 10 MeV to 200 MeV.

The high-energy region is to be explored with another piece of equipment flown in ballons which is shown in Figure 8-5. Again one uses a counter telescope, with scintillation counters, a large gas Cerenkov counter, a sandwich of scintillation counters, and tungsten plates underneath in which the electrons and positrons develop a shower. The number of electrons in each of these layers of scintillator is measured and thereby the shower development curve is obtained; in this way the energy of the electrons can be determined. This equipment was flown last October and was in the air for about thirty hours. At the moment it is at Brookhaven to obtain proton calibration. It is an extremely nice electron spectrometer with which it is hoped to reach energies up to approximately 200 GeV. In the far future an extension up to 1,000 GeV should deliver very relevant information on the storage problem.

# Cosmic Ray Neutron Studies
## SERGE KORFF

*Abstract: The neutron component of cosmic rays provides an important tool to investigate both the equilibrium of the nucleon component in the atmosphere and the fluctuations of the flux incident at the top of the atmosphere. Both these phenomena are surveyed, and a proposed method of long-term geological dating is discussed.*

INTRODUCTION

Although we have been in the neutron business for a good many years, one of the incidental conditions therein is due to our university circumstances. We do not have a very large physics department. This seems very strange to some people because New York University is the largest private university in the world; yet while we have 47 or 48 thousand students (nobody knows exactly how many), the majority of these students are in dentistry, business administration, and all kinds of subjects such as journalism and law which do not involve physics. So we do not have the happy situation such as Berkeley's with twenty-five or thirty full professors and the ability to go to the AEC and say, "Now we need 100 megabucks because we have to build a little gadget off in one corner of our campus!" We have a slightly different set of boundary conditions and we do the best we can within these limitations.

The neutrons turned out to be a great deal of fun, not only because of their own inherent interest but also because they have led into quite a lot of very interesting new avenues which are only just now beginning to open up. One of them is that practically all the neutrons are produced in the atmosphere. They are secondary particles; they are produced by the high-energy cosmic ray primaries, but most of them are produced by protons (some are produced by Professor Meyer's electrons since any entity that comes in with $10^9$ eV is capable of producing neutrons). They diffuse about in the earth's atmosphere, they are slowed down by collisions, eventually they are reduced to low energies, and at that point (with the exception of a very few that diffuse out from the top of the atmosphere) all of them are captured and generate new isotopes. These new isotopes have turned out to be of considerable importance, $C^{14}$ being, of course, the classical example; but there are other isotopes which offer great promise and that will undoubtedly turn out to be important in the future. I'll mention a few more of these later on.

Another very interesting result of the neutron studies is that neutrons are particularly sensitive to solar modulation effects due to flares and other events on the sun which produce effects visible here on earth. In such effects the neutron is not only a good indicator but also it is related to many of the other geophysical parameters such as the strength of the earth's magnetic field, magnetic storms, aurora borealis, ionosphere effects, and many other effects in the earth's atmosphere. These are all interrelated. As a result of the IGY work, we are for the first time beginning to develop a really good picture of what happens when one of these great clouds of plasma is emitted from the sun. Much of what I shall describe comes from a wide variety of sources—the work of many people in addition to our own work. Most of these data are available in places such as the World Data Center. If you are interested in finding out the connections between this and that, the data for many of these analyses exist. Principally necessary is the imagination to figure out how best to put it together.

NEUTRON MEASUREMENTS

The study of neutrons has frequently brought to light some experimental points of considerable interest. Each time we thought we knew what we were doing, some little twist turned up and it wasn't quite the way we thought it was; this is one of the things that makes life interesting and good fun in science. One of the problems is that a neutron detector in the atmosphere may itself behave as a source of some of the neutrons being detected, so one has to be very careful in the design of the experiment to insure that the neutron counter, for example, is far removed from the rest of the apparatus that goes with it. Still another point: a $BF_3$ slow neutron counter will measure not only the neutrons but it will also measure as a background all high-energy events in which some 2 MeV of ionization is liberated inside the counter. So one must think through the problem of how to determine and subtract out the background.

We worked out a long time ago the technique for doing this. Two identical counters are constructed, different only in the degree of isotopic enrichment of boron-10 which is the active isotope (boron-11 does nothing at least as far as this particular experiment is concerned). One of them is filled with the gas enriched in boron-11 and the other is filled with the gas enriched in boron-10 to the same pressure. They must be symmetrically disposed with respect to the batteries and other ancillary equipment so that the secondary radiation generated in the apparatus as well as in the atmosphere has about an equal probability of getting into the two counters. By algebra one can separate the contribution due to the neutrons to give both a background and a neutron determination. When this is applied to neutrons, there is another difficulty: the slow neutron counters look only at the bottom end, the low-energy end of the spectrum.

The neutrons are in fact produced with a whole broad spectrum of initial energies; they are then modulated by being scattered in the earth's atmosphere, and during the scattering process they may be, and some of them are, captured by resonances in nitrogen and other substances. Those neutrons are actually present in the earth's atmosphere, but the $BF_3$ detector does not see them. For this reason one had to also consider whether it was possible to look at the higher energy neutrons, and this cannot be done with a gas counter because of the efficiency argument against it. Use of a scintillation counter produces recoil of protons. By putting this counter into an environment where many high-energy protons exist and in which there is just about any amount of junk in any energy range, one again encounters a background problem. Therefore, one must again devise some combination of subtracting out the background which is one of the parameters measured and is one of the parameters which gives also some additional interesting information with no extra toil.

## NEUTRONS IN EQUILIBRIUM WITH THE ATMOSPHERE

The primary cosmic ray particle, whatever it may be, comes in and hits a nucleus in the atmosphere. It may kick out of this nucleus a fast proton or a fast neutron. The fast protons may hit further nuclei and thereby generate a nucleonic cascade which will eventually lose that energy largely through ionization. There also may be mesons generated. The neutral mesons will make photons and these will generate an electronic cascade. If they have high ener-

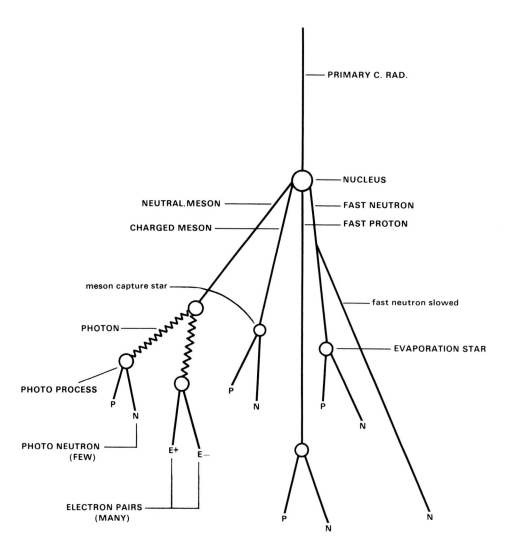

PRODUCTION OF NEUTRONS BY THE COS. RAD.

**Fig. 9-1. Possible neutron production processes by interaction of cosmic rays in the atmosphere.**

gies the charged mesons will generate other nuclear stars and protons and neutrons. A fast neutron may be either degraded by many elastic collisions and simply become a slow neutron, or it may also in turn hit a nucleus and generate still more neutrons by evaporation stars. Altogether, there are many different processes by which neutrons come into the atmosphere as a result of and secondary to the primary radiation. A typical sequence of processes is shown in Figure 9-1.

A kind of balance sheet may be set up as shown in Figure 9-2. A certain

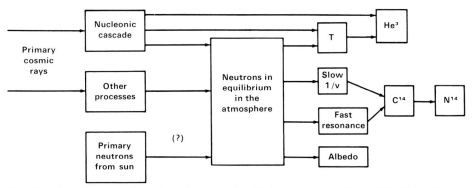

**Fig. 9-2. Flow diagram of neutron "balance sheet" showing production of T, He³, C¹⁴, and N¹⁴ in the atmosphere.**

number of neutrons are at equilibrium in the atmosphere. They come from various sources. Primary cosmic rays come in and they generate a nucleonic cascade which is the probable source of neutrons. There may be, of course, other sources; they can occur as photoneutrons resulting from the high-energy electrons of Professor Meyer or from other high-energy events such as the decay of $\pi^\circ$ mesons. It is possible that there are primary neutrons arriving from the sun. We have spent a good deal of time looking into this problem and the state of our knowledge about it is still represented by vast questions. We must know more about this. It turns out to be not at all a simple problem.

What happens to these neutrons? They are slowed down by the processes of various collisions in the atmosphere; they become $1/v$ neutrons in which case they are likely to be captured by nitrogen to produce $C^{14}$. The $C^{14}$ all eventually decays back into $N^{14}$. There is no accumulation of an isotope in the atmosphere that can be identified. At one time it was thought that slow neutron measurements could get a number match with the $C^{14}$ that Libby *et al.* get from their $C^{14}$ measurements, but we always came out with smaller numbers than his. Finally we realized that there is an additional source of $C^{14}$ from the neutrons captured in the nitrogen resonances while they are still fairly fast which our $BF_3$ counter doesn't see at all. Some neutrons are scattered out of the earth's atmosphere to which the word "albedo" has been applied. It will undoubtedly cause any astronomer to shudder; this is not a proper use of that word. It is also called leakage neutrons, which is perhaps a better term.

Neutrons also generate tritium in the absence of nitrogen, and here again it is not possible to measure the amount of tritium and get the number of neutrons coming from the neutrons in equilibrium in the atmosphere, because there are some neutrons produced directly in the nucleonic cascade. All the tritium in the earth's atmosphere will eventually decay into He³. The measure-

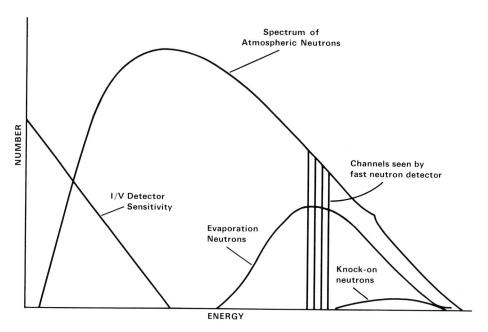

**Fig. 9-3. Sensitivity regions for various detectors superimposed upon schematic neutron energy spectrum.**

ment of $He^3$ turns out to be a very nice useful dating tool because it is stable and is cumulative. It can be shown that all the $He^3$ in the earth's atmosphere at the present time is of cosmic ray origin, and in all probability there was a great deal more $He^3$ that has been generated in the earth's atmosphere but has escaped by the usual escape velocity mechanisms. Again a warning is needed about equilibrium calculations because some He nuclei are generated in nucleonic cascades as part of the spallation processes that happen higher up, so one does not get nice number agreement. There are other processes I have not shown, but other neutrons go into still other nuclear reactions with the atmosphere, with the crust of the earth, and with water.

Some of the problems encountered when one tries to detect neutrons are shown in Figure 9-3. The energy spectrum numbers as a function of neutron energy in the atmosphere is represented roughly by the top curve. This curve rises, has a turnover point, and starts down because in the earth's atmosphere at lower energy levels, the neutrons are taken out by the $1/v$ process of which by far the largest one is the $C^{14}$ process. This turnover is the mission of the $1/v$ $C^{14}$ process. Looking at this kind of system with a $1/v$ detector whose efficiency or sensitivity is represented by the straight line, one is certainly not detecting any neutrons with energies above that amount. However, one can get

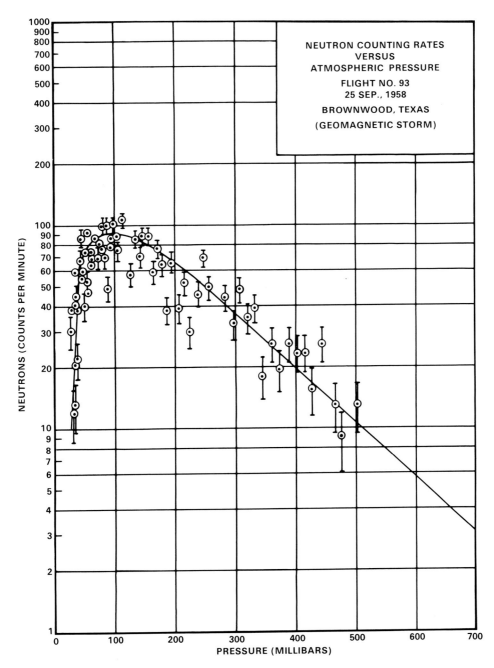

**Fig. 9-4. Typical counting rates in B$^{10}$-enriched counters and normal BF$_3$ counters as a function of depth in the atmosphere.**

a quite nice relationship between the number that are captured in the 1/v process in nitrogen and the number the detector counts. The difficulty is that there are many resonances in nitrogen and in other nuclei also which take out some of the neutrons and which this detector never sees.

There are essentially two kinds of processes which generate the neutrons: a very high-energy neutron may be produced in a dynamic process by a very high-energy primary simply knocking one or two or three neutrons (and other things as well) out of a nucleon. There is no upper limit to this zone; it goes right on up to very high energy. These high-energy neutrons do generate evaporation neutrons because they generate stars and form these lower energy neutrons; so there is a second source which is the evaporation neutrons. One can also look at neutrons using their effects on detectors. In the neutron scintillation counter detectors, one looks at several different channels, neutrons from 1 MeV up to 10 MeV, and breaks them up into channels, and one can determine the number of neutrons of more than 1 MeV and less than 2 MeV, and so forth in each interval. It is a rather unsatisfactory situation because there are no very good detectors that work very well at high energy but there is a plethora that work very well at low energy. We have designed a neutron detector that seems to work at high energy, but it turns out to be rather expensive to build.

The practical means for detecting neutrons in this energy range is to put paraffin around the counter, slow the neutrons down, and then detect them with a $BF_3$ counter. And this is fine provided the geometry of the experiment will permit it. In front of a cyclotron everything can be set, the geometry is under control, and all is well. For cosmic rays, one does indeed detect the neutrons, but also other neutrons which are manufactured in the paraffin, and two or probably three variables are measured in a single experiment. There is no very good way to determine how much is attributable to each of these sources. Experiments, which several people have made using paraffin wax and other material around the counter, give numbers that are very hard to interpret. To find the numbers in equilibrium with the atmosphere and to solve what could be done to avoid the additional complications from locally generated neutrons is rather thankless business.

Figure 9-4 shows the data for an enriched and a depleted counter on a balloon flight from Brownwood, Texas. The neutron counting rate is plotted against atmospheric depth in millibars. The top of the atmosphere is at the left and sea level is off the scale to the right. There is the usual transition maximum near the top of the atmosphere. By doing algebra on these two curves one can obtain the number of neutrons.

After the algebra is done on the data from two counters one can determine the background. A typical result is shown in Figure 9-5. This is not neutrons.

The background of high-energy events in one of these BF$_3$ counters continuously increases with altitude instead of having a nice turnover point and going down. At the high elevation there are many high-energy events in which a

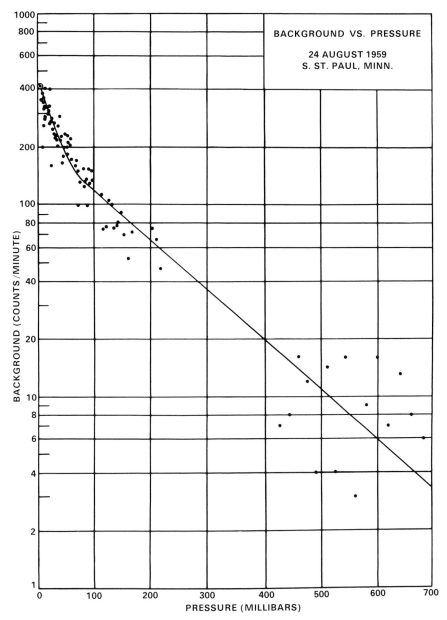

Fig. 9-5. Typical non-neutron background experienced by both counters in the pair as a function of atmospheric depth. Deduced from data of the kind shown in Figure 9-4.

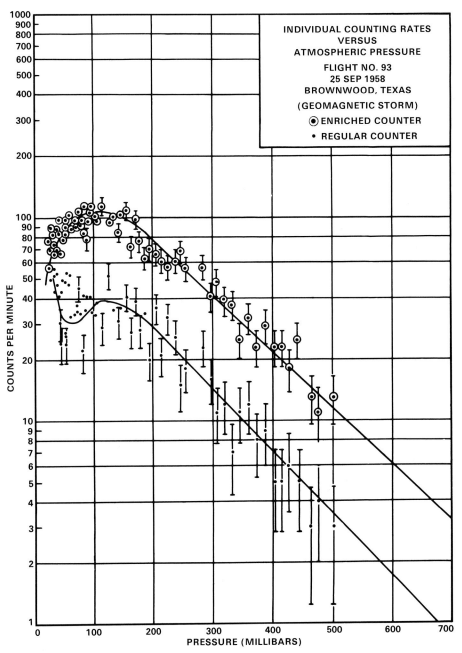

**Fig. 9-6. Typical slow neutron spectrum obtained from a pair of BF₃ counters. Deduced from data of the kind shown in Figure 9-4.**

good deal of ionization is generated in the counter. The fact that this curve turns and goes up has been a great pest in analyzing the data. By contrast the

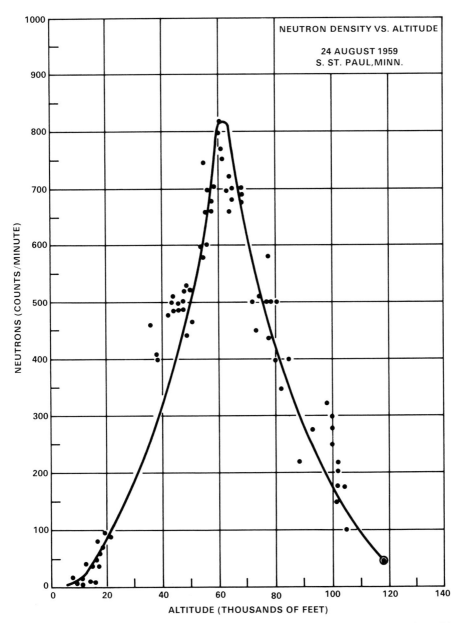

**Fig. 9-7. If the neutron spectrum of the type shown in Figure 9-6 is plotted against altitude rather than atmospheric depth in millibars there is an apparent remarkable change in shape.**

number of neutrons shows a maximum and then it comes down as in Figure 9-6. One of the flights climbed to within 4 gm of the top of the atmosphere. These are the $BF_3$ neutrons, very slow neutrons. The number of very slow neutrons at the top of the atmosphere is extremely small. Extrapolation of the curve to intersect the ordinate would give the leakage flux of very slow neutrons having actually escaped upward from the top of the atmosphere. Data for this same experiment with a fast neutron counter show much larger numbers at the top of the atmosphere so a bigger fraction of the fast neutrons are going out.

It is interesting to observe how the apparent shape of the neutron curve changes if neutron numbers are plotted against a uniform altitude scale instead of pressure (which intrinsically contains the logarithmic term of the exponential atmosphere). This can be seen in Figure 9-7 where the number of neutrons has a maximum which is almost symmetrical, except on the right it will flatten out eventually at the top of the atmosphere for the albedo, the neutrons being scattered out of the top of the atmosphere. The curve looks amazingly different when one sees it on this other scale although Figures 9-6 and 9-7 are identical. In order to investigate the flattening at high altitudes we sent a set of the enriched and depleted counters to an altitude of a couple of hundred miles in an Aerobee rocket shown in Figure 9-8 because we wanted to see if our analy-

**Fig. 9-8. Final instrumentation adjustment on Aerobee 4-16 sent to an altitude of 200 miles to measure neutron flux.**

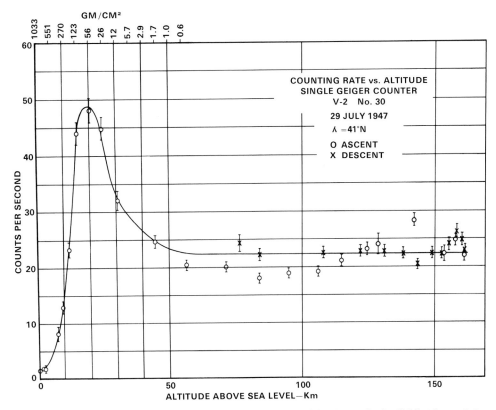

**Fig. 9-9.** The shape of the counting rate of ionizing particles in a single G-M tube out to about 100 miles above sea level is very similar to the shape of the neutron curve. (After Van Allen.)

sis was somewhere near right. We did get a curve which shows a flattening out above the atmosphere. It is very similar to the kind of curve Van Allen obtained for the counting rate of ionizing particles in Figure 9-9. Starting from sea level, it reaches a maximum, then flattens out at about 50 km; thus in terms of the ionizing radiation 50 km is the top of the atmosphere. Note that the peaks in Figures 9-9 and 9-7 occur at the same altitude.

NEUTRON FLUCTUATIONS

It has become clear that events on the sun generate radiation which increases the number of neutrons in the atmosphere. Some solar flares are events of this kind. Figure 9-10 shows a portion of the sun near its limb. One sees a sunspot pair and an active flare region, and also some loops of a loop prominence. Once in a while as a result of a flare of this sort, solar material is suddenly

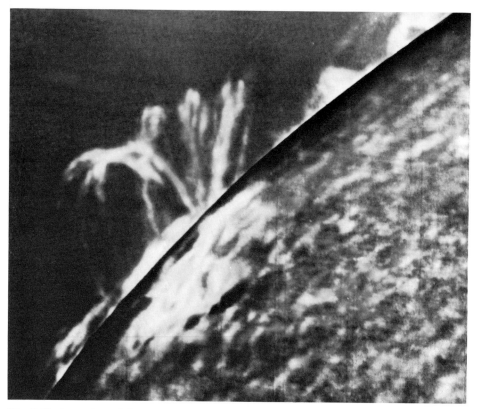

Fig. 9-10. A portion of the sun near its limb. A loop prominence rises high above a sunspot pair and an active solar flare region.

blown out into space and appears as a great cloud of plasma which in the proper set of geometrical circumstances may envelop the earth and produce some of the effects now to be discussed.

Figure 9-11 is a spectroheliogram from the McMath-Hulbert people showing the large flare of 16 July 1959. This flare produced cosmic ray effects visible here on earth. This is a normal spectroheliogram of the sun with the usual configuration of two active bands, one in the northern and one in the southern hemispheres, spot pairs, and a rather large flare which generated particles that came right through the earth's atmosphere and reached sea level.

In November 1960 a region on the surface of the sun gave rise to several spits of radiation in succession. Figure 9-12 shows how they affected the neutron monitor running at College, Alaska. On November 12 there was a large event with a double peak. In the meantime a Forbush decrease occurred so that on November 13 and 14 the intensity was below normal. On November 15 a second shock of neutrons was generated in our upper atmosphere

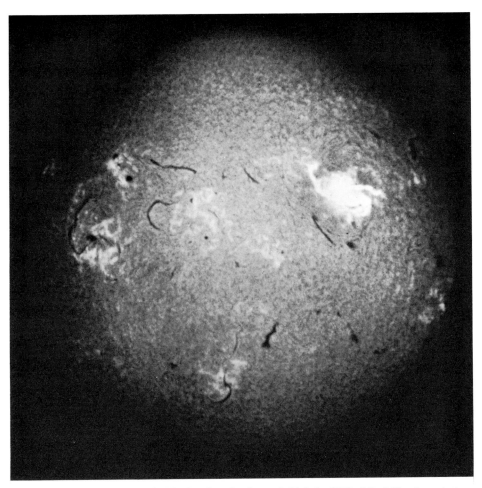

**Fig. 9-11. Spectroheliogram in H₂ showing the flare of 16 July 1959, 21ʰ 54ᵐ universal time. It produced cosmic ray fluctuations. (McMath-Hulbert Observatory.)**

and then they tended to return slowly to normal with the exception of an interesting event on November 20 which I shall discuss later.

Why does one monitor the neutron component? Why not use the ionizing component of the cosmic radiation? The answer is, of course, both are done. The Berkeley group had all three detectors going at the time of the big flare February 1956 (see Figure 10-18); they had a meson telescope measuring the hard component which saw this with an increase of about 30 per cent in amplitude. They had a total ionization measuring device which recorded an increase of 50 per cent from 600 to 900. But their measurement in the neutron component showed an increase of from 30 to about 210, a factor of 7. It is because of this enormously greater amplitude of effects in the neutron compo-

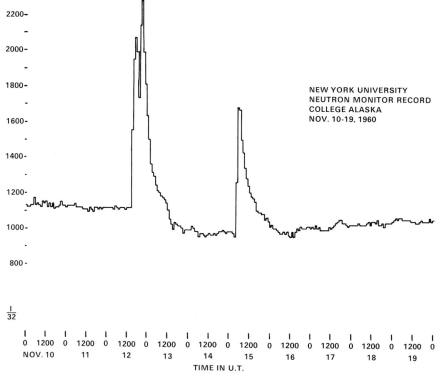

**Fig. 9-12. Between 11 and 19 November 1960, several events on the sun caused marked changes in the neutron flux in Alaska. A double peak, a Forbush decrease, and a second peak are visible. (See also the event of 20 November in Figure 9-14.)**

nent that the neutrons have been considered a very good indicator of this phenomenon and very useful in enabling one to learn a little more about it.

Additional important information about these events is obtained by looking at the changes in other geophysical parameters which occurred at the time of this event as shown in Figure 9-13. One can see the neutron curve with a double peak and the later peak on November 15. A meson telescope operated by Professor Chasson in Lincoln showed a Forbush decrease which occurred about six hours after this event, followed by the usual recovery back to normal. There was a sudden commencement magnetic storm which was signaled at precisely this instant; there was the onset of a flare of type $3+$ on the sun; the magnetic $K_p$ index went up very rapidly and then eventually came back down again. Much of this pattern was repeated on November 15. There was a radio type-IV burst which occurred precisely at the instant of the onset of this flare with a sudden commencement and also with the start of the increase of the cosmic ray activity. The aurora borealis was visible in New York that night, ordinarily far south of the aurora zone; it was not coincident with

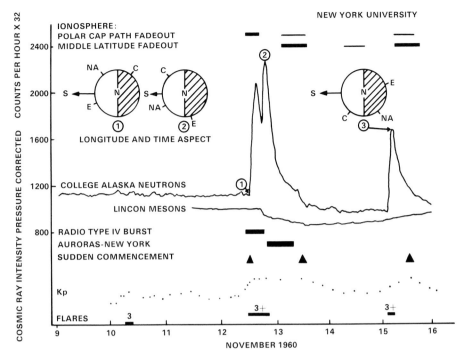

Fig. 9-13. Events of 12 through 15 November 1960 are correlated with: polar cap path fadeout, middle latitude fadeout, longitude aspect of earth to sun, neutron monitor data, meson monitor, radio type-IV burst, aurora borealis, sudden commencement, $K_p$ magnetic index, and solar flares.

the start but approximately coincident with the Forbush decrease six or eight hours later. Data from Dana Bailey at the Radio Laboratory in Boulder, Colorado, showed that exactly coincident with this there was a polar cap fadeout event in which the transmission paths over the North Pole region simply went to zero; twenty-four hours later, when the polar cap fadeout was beginning to recover to normal, there was a middle latitude fadeout in which the middle latitude transmission paths went down. There are time-lag effects in the ionosphere. That pattern was also repeated on November 15.

Can these events be correlated with the aspect of the sun and the earth at the time of these various fluctuations? The positions of the earth as viewed from above the north geographic pole are superimposed upon Figure 9-13 and are keyed into the College, Alaska, neutron data by numbers 1, 2, and 3. The direction to the sun is shown by the vector S, NA is North America, C is College, Alaska, and E is Europe. At time 1 the particles arriving at College had to come around the curvature of the earth. At time 2 College, Alaska, had moved into the daylight hemisphere and Europe had already gone into the night (European observations confirmed this), so again the particles came

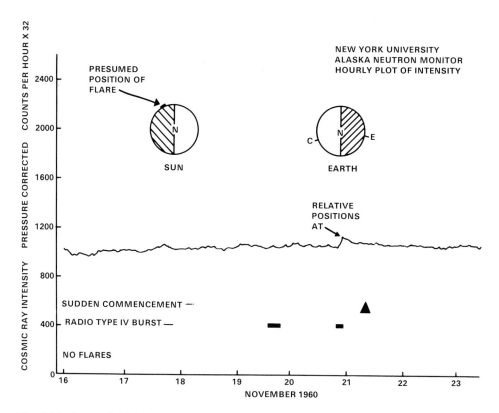

**Fig. 9-14. Event of 20 November 1960, correlated with data similar to Figure 9-13 except no solar flare was visible on our side of the sun.**

round to the backside of the earth. At time 3 both North America and Europe are on the dark side of the earth but College is still in daylight.

Turning now to the event of November 20, a very definite increase in cosmic radiation was observed at College. At that particular time the same region on the sun which had given those two increases on November 10 and 15 had rotated around so it was behind. There seems to be no doubt at all that another big flare occurred and some particles came around from the backside of the sun as shown in Figure 9-14. This event was seen in Europe as well as by the monitor in College. No flares were reported, there was a radio type-IV burst, and there was also a sudden commencement reported at this time. Until we saw events like these, some of us had not realized that we could get effects on the backside of the sun which would be "visible" here at the earth.

A model for the Forbush decrease is illustrated in Figure 9-15. In phase 1, one assumes that a region on the sun directly in line with the earth emits a plasma cloud which may or may not be connected with lines of flux going back to that point; this would correspond to the undiminished intensity previous to

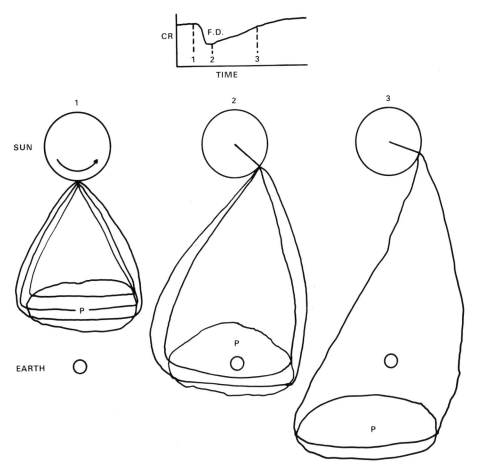

**Fig. 9-15. A model to explain a Forbush decrease. (See text for explanation.)**

the arrival of the plasma cloud. In phase 2 the cloud has advanced and covers the earth, which makes it a little more difficult for particles from great distances to diffuse into this; thus, the total number of particles arriving here at earth is a little less. In phase 3 the cloud has passed the earth and goes on out into space and eventually recovery takes place.

SUMMARY

I have discussed some of the other lines of research that cosmic ray studies lead into, as well as some of the features of the neutron studies. A great deal more can be done in the basic problem of the geophysical connections with the other parameters. In fact, the whole subject is in its infancy. There is a great deal of work to be done to find isotopes and their use, which isotopes in turn have been generated by the cosmic ray neutrons. It is a broad and exciting field of study.

DISCUSSION

(Q) One of the interesting questions is how long we have been on the earth. It seems to me you are saying there are isotopes produced by cosmic rays which can accumulate with time.

(A) Korff: There are two isotopes that accumulate. One of them is $He^3$ which is a stable isotope and which, if you can show that the amount is right to have been of cosmic ray origin, can be used for dating purposes because it is cumulative. The second one about which I am immensely hopeful is $Be^{10}$ because it has a 2.5 x $10^6$ year half-life. It is known to exist (Peters has identified it). It accumulates in ocean bottom sediments. A number of other people have worked on it, but nobody has yet done the job on $Be^{10}$ that Libby did on $C^{14}$. It is a safe prediction that if someone does the same kind of work on $Be^{10}$ to make it as useful to geology as $C^{14}$ has been, it will deserve a trip to Stockholm. A 2.5 x $10^6$ year dating tool would be an absolutely marvelous thing. It could be used for the Pleistocene and date whole geological eras with some precision.

(Q) It is my understanding that it is now possible to predict the onset of solar flares.

(A) Korff: Unfortunately the situation is not good. Much work has been done, but how well we can predict a solar flare is not in a very happy and satisfactory state just now.

(Q) What are some of the models which have promise to account for the origin of solar neutrons?

(A) Korff: At the time of a solar flare if particles of energies up to 15 GeV or more are generated, it is quite certain that some neutrons will be generated at the same time, if only by charge exchange scattering. If there is any accelerating mechanism that makes high-energy-charged particles on the sun, then by any one of a number of reactions there will also be neutrons. Above the threshold for neutrons there will be neutrons by collisions and all the other multitudinous processes which make them. If acceleration of the very high-energy particles on the sun takes place very far up above the photosphere, the densities there may be so low that not many neutrons per unit volume will be produced, but even so, there should be some. There is also the problem of transit time from the sun, the sun being 8 light-minutes away, a time that is comparable to the lifetime of neutrons in free space. Only the neutrons well up toward relativistic energies will arrive at the earth. A simple calculation shows that the number of neutrons at 10 MeV which survive to reach the earth is only a few per cent. At relativisistic energies many neutrons would certainly arrive at the earth. Up to the present we have not had the good luck of having a neutron detector above the earth when one of these big flares occurs. Nowadays there are many neutron detectors flapping around in satellites and there are various proposals for airplane and balloon flights. The difficulty with the balloon flights is that by the time it is known that a flare exists it is too late to rush out to inflate the balloon, then wait two hours for it to reach the top of the atmosphere. A similar problem exists with airplanes. Seldom is there the possibility of a standby airplane flight. Flares produce relativistic particles which arrive here at nearly the same time as the optical signal does. In other words, the 15 GeV are coming at very nearly the speed of light. Should there be 15 GeV neutrons you must look for them the same instant. In other words, the detectors must be already up there. It is a very interesting and intriguing problem. I should be very much astonished if there were no solar neutrons, but I would also expect that they would be connected with disturbances on the sun. Under normal circumstances there would be very few neutrons, but you might get shots of them when these events occurred on the sun in which high-energy particles are also generated.

# Neutron Exposure in Supersonic Transport ROGER WALLACE

*Abstract: The exposure to cosmic radiation experienced by passengers and crew in a supersonic transport at 65,000–75,000 ft has been investigated. The problem is very complex due to the different dependences of the various cosmic ray components upon altitude, the differences in tissue response to radiation, the large variations in cosmic ray intensity, and the effects of the earth's magnetic field. It appears that other supersonic transport problems are more serious than those of radiation exposure.*

INTRODUCTION

The previous speaker's suggestion that we have $10^8$ available at Berkeley when others have less is too high by three or four orders of magnitude, especially in the work I shall describe. The supersonic transport (SST) is one of the technical developments that is ahead of us in this country. It has many unsolved problems which have been ignored. I believe the attraction to satellite work has been responsible. There are experiments which should be done to check out many different features of the SST's. This is all the more urgent because no military prototype exists to guide us. Of course, work like ours starts with a typical proposal sent to the federal government to show why the SST is valuable and that the company should study the problems which cosmic rays present to that transport. Figure 10-1 shows the cover of the proposal.

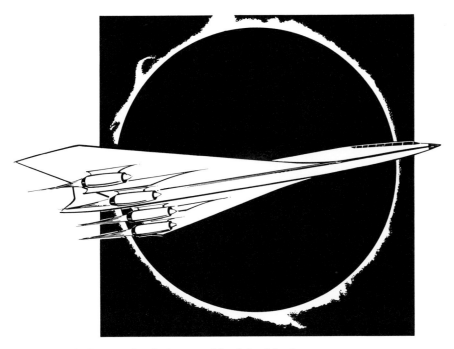

**Fig. 10-1. Typical cover of typical proposal for federal funds.**

BASIC PROBLEMS

There are five basic problems in connection with the development of the SST. They are: (1) cost (construction vs purchase trade balance), (2) heat (aluminum vs steel and titanium), (3) fuel and jets, (4) sonic boom, (5) radiation. I shall touch on the first four briefly, then discuss the last one in greater detail.

The cost is a very serious problem. Basically, if we do not build the supersonic transport (and many critics have said we should not) we will have a $20 billion per year adverse trade balance, because within a year the Russians, the French, and the British will have their versions of a somewhat different supersonic transport flying, and we shall be forced to buy them.

The heat problem is yet to be solved. No non-military aircraft has ever flown for any reasonable period, and even military aircraft have not flown for more than an hour or so, at the altitudes and speeds that the SST must attain. It is proposed that the SST will cruise at 65,000–75,000 ft and will develop skin temperatures of about 400 to 600° F. At Mach 2.5 or so, which is the speed of the British, French, and Russian aircraft, the skin temperature is about 400°F. and they can use aluminum with conventional aircraft construction. At the Mach 3 speeds of the American aircraft steel or titanium must be used because the temperature goes to 600° F. This actually introduces many manufacturing simplifications because steel is far easier than aluminum to fabricate, but it also represents a new technology and a very expensive one so that the advantages of steel are offset by the newness of this product in the aircraft industry.

The fuel problem is almost never mentioned in the news media. No jet aircraft has ever flown using the fuel that is intended for SST use. The fuel is not conventional jet kerosene, but it is to be a material which changes state as it is warmed up. This change of state is to be used to absorb heat from the aircraft skin, and heat exchangers will be provided to cool the air circulated inside the skin to take the 600° F. down to some reasonable temperature for the passengers. There will be heat exchanged with a non-inflammable intermediate fluid and that heat in turn exchanged with the fuel. Such a system is obviously expensive, complicated, and fraught with difficulties. Also at present the supply of fuel is very small; it exists only on a laboratory scale.

The sonic boom problem has received more public attention than any of our other problems. It has been tested in the only state in the United States where more than half the population is supported by aircraft salaries, and there were quite a few complaints in spite of that. It may not be a serious problem because if one assumes that all supersonic flights will be for distances of over 3,000 miles, it follows that most of them must be over the oceans. There has never been a restriction on sonic booms over the ocean and no complaint has ever been received in writing from any ship or from anyone on any island; apparently the public relations over water are satisfactory and one can have any kind of sonic boom one may produce. On the other hand, many complaints have been received about sonic booms over land. It is possible to arrange flight paths that do not disturb too many people, and it is also possible that the sonic boom may not be as serious as one thinks. In some cases there is good evidence that the average population cannot actually tell the difference between a supersonic aircraft and a conventional aircraft—they just do not like the noise. The advantage in the neighborhood of the airport is that the supersonic aircraft with such enormous surplus power can climb away from the residential areas more rapidly, resulting in less noise around the airport than at present. This is shown in Figure 10-2 where it can be seen that the SST height at 3 miles downrange from the starting point is something like twice as much as the subsonic jet. Another favorable factor is that the noise pattern is directed in a very small cone forward from the nose because of the way the jets are enclosed under the tail. This is compared with the typical compressor whine of a conventional jet in Figure 10-3. Of course, there will be no supersonic flight at altitudes below about 40,000 ft, so there is no question of sonic boom around airports—just the compressor whine which all have heard many times. Thus the SST will be quieter around airports.

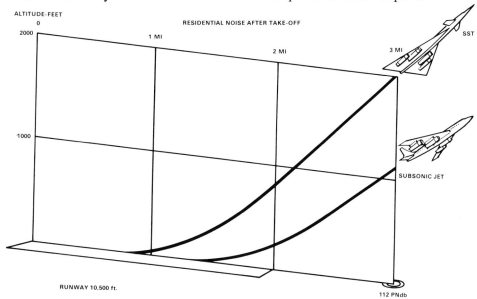

**Fig. 10-2. Less residential noise after takeoff is expected from the SST (supersonic transports) than from the conventional subsonic jets because of faster climb rate.**

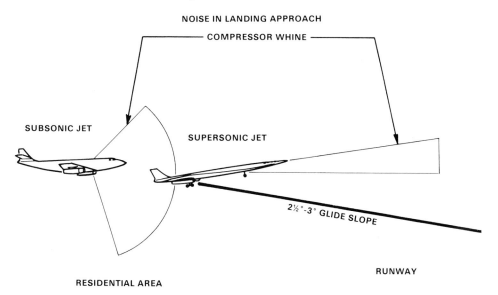

**Fig. 10-3. The SST will have less noise in landing approach because it has a smaller cone of compressor whine.**

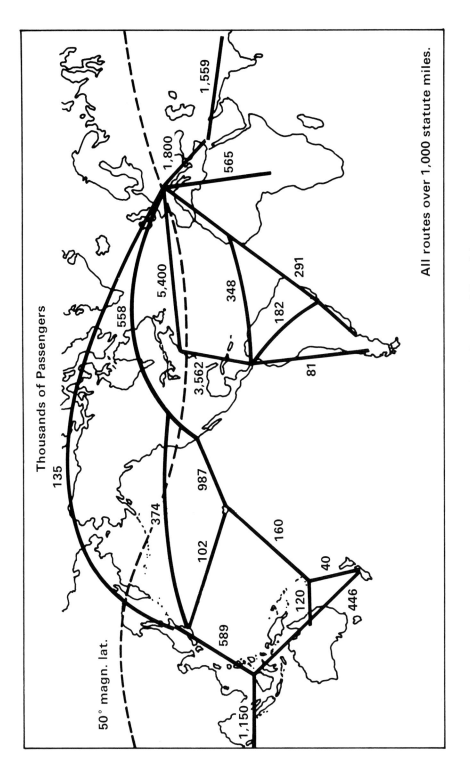

**Fig. 10-4. World airline traffic flow. (Courtesy of Convair.)**

RADIATION EFFECTS

Although radiation is a quiet problem and it may turn out that it is not a serious question, it is the hazard of greatest public concern. It is also the problem that has received the highest level of technical effort comparable to various other questions. (Examples are the present unavailability of aircraft instruments for an airplane such as the SST, and various other problems connected with pilot visibility on landing which have strong effect on the real safety of the aircraft but are not very interesting to most people. Perhaps people are prone to worry about the public problems such as radiation and sonic booms, and they do not realize the many unsolved technical problems of this aircraft.) The radiation difficulties are nonetheless real. It is well known that the cosmic rays are influenced by the earth's magnetic field which causes them to arrive at the earth with substantially no energy selection in the area of the magnetic poles and to be filtered by the magnetic field in the equatorial regions. Through accidents of geography and history, the major long-distance air routes are inside the United States, the United States is nearer to the south magnetic pole, and our magnetic latitudes tend to be higher than that of most any other area in the world. It turns out, however, that the routes which are mainly affected by this low magnetic screening are the transatlantic routes from New York to Europe, from the West Coast to Europe, and the rather sparsely traveled polar routes from Alaska and Japan to Europe. World airline traffic flow is shown in Figure 10-4. Although these numbers are about ten years old, the relative importance of the routes has changed little, and the real concerns are the New York–to–Europe route and the West Coast–to–Europe route. If the radiation situation is satisfactory on those routes, all of the other routes will take care of themselves for magnetic reasons. In the relation between geomagnetic and geographic coordinates, the United States is at a higher latitude because the magnetic equator is depressed in the South American region far south of the geographic equator. This means that the south magnetic pole which is in the northern hemisphere is quite a bit in our direction from the geographic pole and therefore we have an unusually high magnetic latitude. Some great circle routes are plotted here.

Total passenger figures do not give the complete picture. Table 15 shows a breakdown of the airline passenger figures. The genetic implications cannot be ignored (more of this later). Note that the breeding age is eighteen to twenty-eight years. People in the older brackets, who are not breeding, represent the major part of the flying population. The SST will further select for old age, because there will be a surcharge on the passenger ticket, and there will also be a competing air bus service with a reduced rate. Only the very affluent

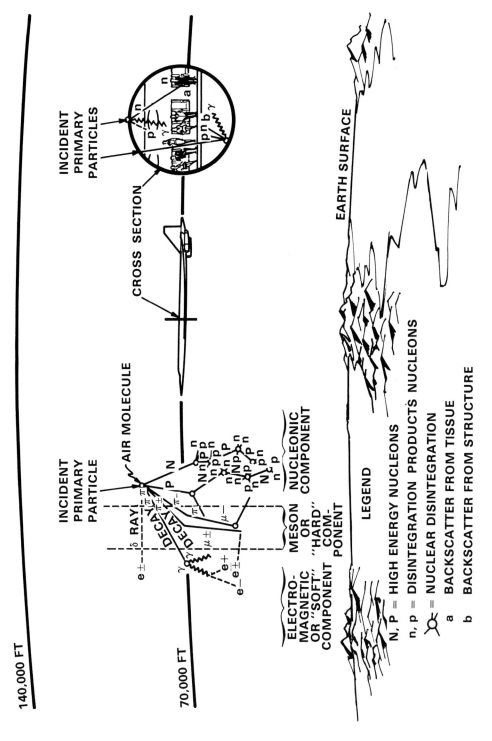

**Fig. 10-5. Development of typical cosmic ray secondary particles in the atmosphere and in the SST aircraft.**

**Table 15. Age and Sex Distribution of Eastbound Transatlantic Passengers, as Compared with the Population of the United States.**

| Age (Yr.) | Passengers (a) | | U. S. Population (b) | |
|---|---|---|---|---|
| | Male % | Female % | Male % | Female % |
| Under 12 | 5.9 | | 25.3 | |
| 12-19 | 3.7 | 5.5 | 7.0 | 6.8 |
| 20-34 | 14.0 | 14.0 | 9.2 | 9.3 |
| 35-64 | 28.1 | 21.4 | 16.2 | 16.9 |
| Over 65 | 3.7 | 3.7 | 4.1 | 5.2 |

(a) Total passengers: 610,000. Between April and September 1963, approximately 610,000 eastbound transatlantic passengers were estimated to travel on flights leaving New York (A. J. Raebeck, Port of New York Authority, personal communication.)

(b) Total population: 186,591,000 (based on U. S. Census, 1963).

will be traveling by SST and they are very likely to be over the breeding age. Therefore, the pessimistic picture from magnetic latitudes will be vastly altered; only a very small part of the genetic pool will be exposed to radiation in the SST—probably less than 10 per cent. This number changes a little because when the SST was projected, the air bus (which will probably come into production much sooner) was not yet proposed.

The basic physics of the SST radiation problem is shown in Figure 10-5. The incoming primary cosmic ray (most probably a proton) interacts in the upper atmosphere. There are 1031 gm of air between the top of the atmosphere (wherever that may be) and the sea level. Half of the air (500 gm) is below 18,000 ft. The SST will probably achieve a supersonic velocity at about 40,000 to 50,000 ft, will cruise starting at 65,000 ft, and as its fuel is used up it will drift up to about 75,000 ft so the radiation must be examined between 65,000 and 75,000 ft. This corresponds to about 60 to 70 gm of air above the craft so that it is under a small part of the atmosphere—less than 10 per cent of it. This is in a region where the primary particles are mostly converted to secondaries but not entirely. There are many different components—neutrons, protons, pions, various types of electromagnetic radiation, and tertiary particles which have been studied in detail. The aircraft, of course, will not be able to carry any shielding because of the weight even though shielding might be desirable.

Thus the problem is that under something like 60 or 70 gm of air one has a few grams of fuselage, a little bit more air, then something like 10 to 20 cm of tissue, and one wants to determine the dose inside of this tissue. The principal concerns are the blood-forming organs, the ovaries, the testicles, and the

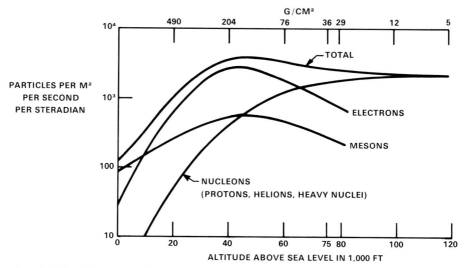

**Fig. 10-6. Transition curves for several cosmic ray components.**

skin. Since the radiation has a rather complicated spectrum, is made up of many different particles, and penetrates different thicknesses of shielding and tissue above these various organs, it becomes quite a complicated radiobiological problem to determine the radiation dosage and what it will mean to the individuals in flight. Our present work indicates a very minor radiation dose compared to the dose that the average person gets from medical x-rays. Nevertheless, one must obtain a quantitative result to reassure both researchers and the public, so we shall assemble the facts about the radiation and its biological effects.

RADIATION DOSAGES

The radiation source in which the SST is immersed is represented by the transition curves in Figure 10-6, in which particles $m^{-2}$ $ster^{-1}$ $sec^{-1}$ vs altitude are shown. The total number of particles increases with altitude until in the region of 40,000 ft it starts to decrease, with a very broad peak. Assuming the SST flies at about 70,000 ft, the nucleons are still increasing, the electrons are decreasing, and the mesons have begun to decline. Since the mesons and electrons are not very effective biologically the main problem is with protons and neutrons. The heavy nuclei constitute a very small fraction of the total. The neutron and proton dosages are the principal ones.

The various parts of the body afford vastly different shielding effects, as shown in Figure 10-7. The skin, with no shielding at all, is a sensitive organ

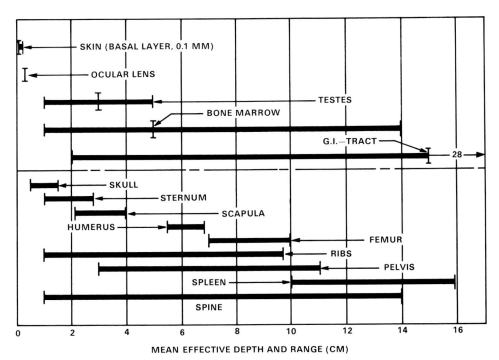

**Fig. 10-7. Depth-distributions of various critical organs and tissues and their assumed mean effective depths.**

in the sense of being available to the radiation, but fortunately the skin has been irradiated for a couple of billion years and man has developed significant resistance so that the biological resistance of the skin to radiation is three or four times that of the internal organs. Thus the skin is not a serious problem in the SST, but it probably is a controlling problem in the astronaut situation. The lens of the eye is poorly shielded and is fairly sensitive, so one must investigate this for the SST crew. Other problems arise. The gastrointestinal tract is the traditional problem for massive doses. The bone marrow must be examined for long-term effects. It is distributed in various bones at various depths, but on the average it occurs at depths of about 5 gm for most of the internal organs, so that little additional material is added to the 60 or 70 gm of the atmosphere. Thus if one makes an estimate of the dose in the free air, one is not too far off for most of the organs in the body.

The effect of the earth's magnetic field upon the total nucleon flux and upon four constituents of this flux is illustrated in Figure 10-8. The most significant behavior is the geomagnetic cutoff for different particles at different geomagnetic latitudes. At 30° magnetic latitude, we receive only that part of the spectrum to the right. At 55° the cutoff allows all energies above that point to arrive at earth. At 90° particles of all energies can arrive. At the geomagnetic equator the cutoff is at 15 BeV/nucleon. Thus, at latitudes below 55° magnetic, a large fraction of the spectrum is screened out; note this is a logarithmic scale.

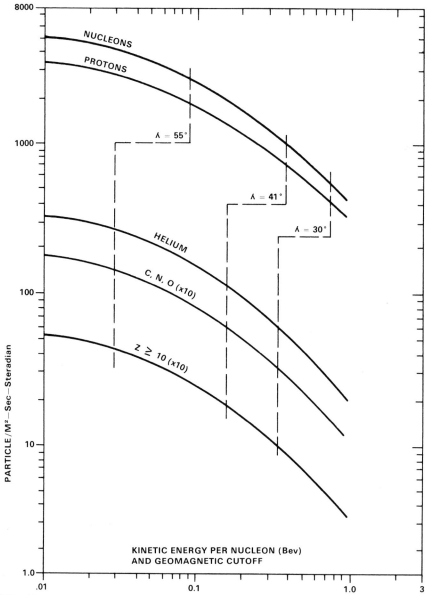

**Fig. 10-8. Integral energy spectrum of primary cosmic rays, separated into four constituents: nucleons as a whole; protons; helium; carbon, nitrogen, and oxygen, and Z = 10. The magnetic cutoffs for 30, 41, and 55 degrees geomagnetic latitude are shown. The cutoff at the equator for protons is 15 billion electron volts.**

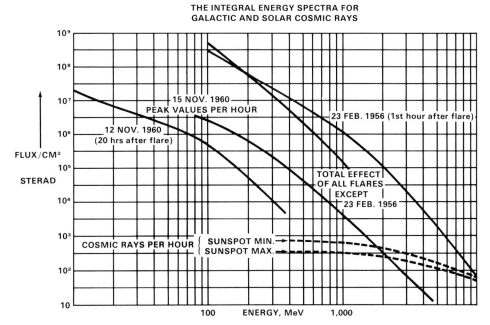

**Fig. 10-9. The flux of solar flare particles is many orders of magnitude above the steady flux of galactic cosmic radiation. Note that the latter is relatively independent of the sunspot cycle.**

SOLAR EFFECTS

The galactic cosmic rays have a rather constant flux, a very, very flat spectrum, and a fairly low intensity which depends only weakly upon the sunspot cycle. The solar cosmic rays, on the other hand, have much steeper spectra which change by many orders of magnitude in flux. The historic solar flare of 23 February 1956, which initiated our great concern about flares, had a very steep spectrum at fairly high energies. These effects are shown in Figure 10-9. The important question arises in the region above 50 MeV where the flare spectra are rather intense and shielding is impractical. Certainly above 100 MeV shielding is out of the question because the mean free path for a 150 MeV particle is about 18 in. of concrete which cannot be incorporated into any airframe design.

An important relation is the dose as a function of atmospheric depth, as shown in Figure 10-10. The SST altitude of 65,000 ft is at about 60 gm cm$^{-2}$. Galactic cosmic ray intensities for several different parts of the solar cycle are shown. The solar cycle is at a peak at this time so that there is a minimum galactic intensity. The galactic intensity changes by a factor of two. The solar cosmic ray spectrum is extremely steep, and any attempt to estimate the effect

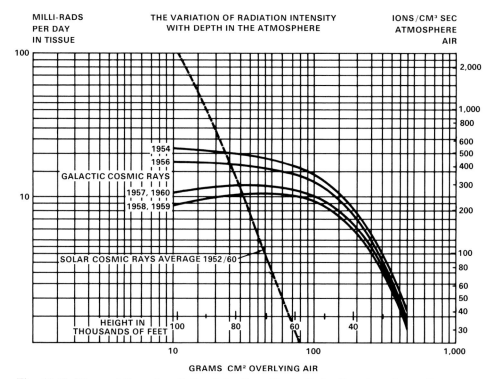

MILLI-RADS PER DAY IN TISSUE

THE VARIATION OF RADIATION INTENSITY WITH DEPTH IN THE ATMOSPHERE

IONS/CM³ SEC ATMOSPHERE AIR

GRAMS CM² OVERLYING AIR

Fig. 10-10. **The variation of radiation intensity with depth in the atmosphere. Curves are plotted for galactic cosmic rays for various years from the last sunspot minimum 1954 up to the sunspot maximum in 1958-59. They are replotted from the work of Neher and Anderson (1962). Also plotted for comparison is the solar cosmic ray component averaged over the nine years 1952-60, as computed for this report.**

of this intensity reveals it as so steep that very small changes of altitude make a big difference in the intensity. Note the ordinate is in intensity units for tissue. In Figure 10-11 are plots of the dose versus altitude. The doses are seen to be insignificant compared with a chest x-ray of about 50 millirems. One would encounter such a dose only at very high altitudes.

The solar flare cosmic radiation presents a different kind of problem. In Figure 10-12 we see the flares in the last ten years or so which produced radiation that could be measured at these various altitudes. These have not actually been measured at 65,000 ft; these are all calculations interpolated from other altitudes. We see the flares are random and unpredictable, but only two of them exceeded the 50 millirem/hr level and those only slightly—the large 1956 flare and another one in 1960. This is equivalent to one chest x-ray per hour, and two hours would be a long flight. The passenger would receive the equivalent of about two chest x-rays if the flare remained at its full intensity for the entire flight of two hours. It is conceivable that once every hundred years a much larger flare might occur. In that case, some sort of evasive action might be considered, but judging by the flares measured it is probably unnecessary. No measurements have actually been made at this altitude during a real flare, because an aircraft cannot get into the air before the flare has died down (due in no small measure to the inordinate delay in paper work to authorize such a flight). Furthermore, at present all aircraft capable of flying at this altitude (mainly the U-2's) are being used for the clear air turbulence (CAT) experiments. This very serious problem for jet aircraft has caused several crashes in the past few years, and these airplanes are flying over conventional routes at these altitudes (and lower) in an effort to devise a method

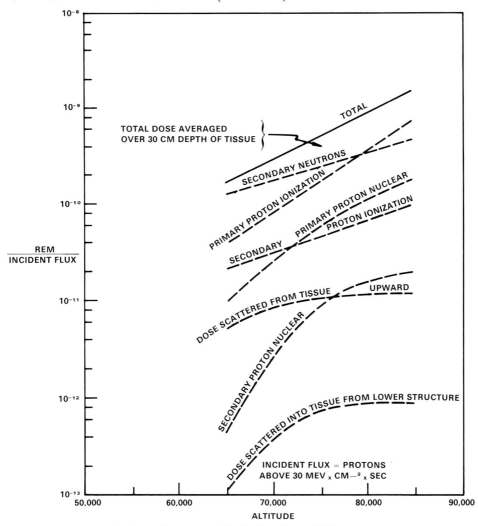

**Fig. 10-11. Dose per incident flux versus altitude (for $P_o = 100$).**

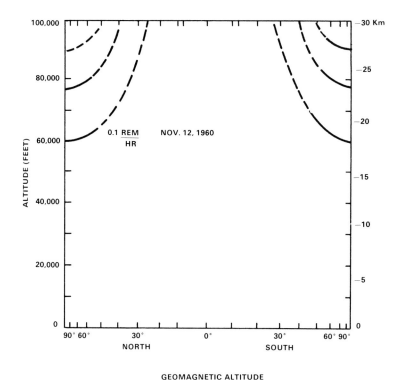

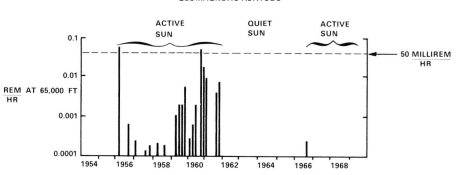

**Fig. 10-12. Solar cosmic ray isodose lines as a function of altitude and latitude.**

to avoid CAT. This is considered a much more important project than the radiation problem.

Sunspot variations have been observed for a very long time. Figure 10-13 is a famous summary of sunspot numbers since 1700 A. D. Although data are scarce for the years before 1700, it is interesting to note that the latest solar cycle which caused all this excitement was the largest solar cycle in something over six hundred years. Long-term periodicities have been studied, and it is predicted that we will not again get such an intense solar cycle for several hundred more years.

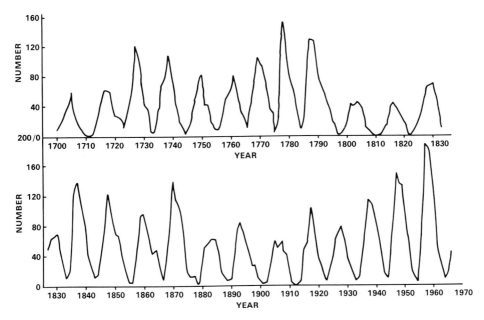

**Fig. 10-13. Time variation of mean annual sunpot number.**

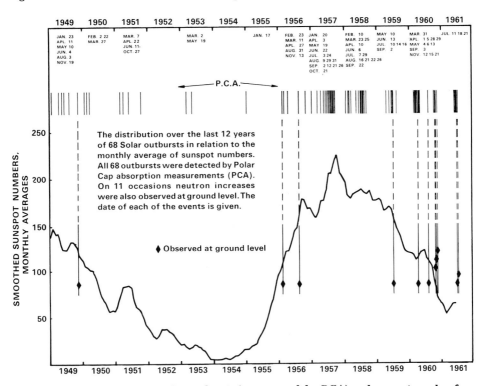

The distribution over the last 12 years of 68 Solar outbursts in relation to the monthly average of sunspot numbers. All 68 outbursts were detected by Polar Cap absorption measurements (PCA). On 11 occasions neutron increases were also observed at ground level. The date of each of the events is given.

♦ Observed at ground level

**Fig. 10-14. Neutron increases, solar outbursts (as measured by PCA) and sunspot number from 1949 through 1961. Note the close correlation among them.**

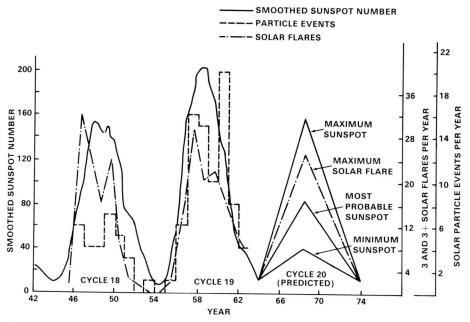

**Fig. 10-15.** Predicted limits and most probable value of sunspot number, and predicted maximum number of solar flares between 1964 and 1974, based upon statistical analysis of the two previous solar cycles.

**Table 16. Maximum Hourly Dose Rates at the Geomagnetic Poles during the Giant Flare of February 23, 1956, in Relation to Altitude.**

| Altitude (Ft) | Dose Rate (Rads/Hr) | | |
|---|---|---|---|
| | Present Estimates | | Earlier Estimates |
| | Maximum Primary plus Secondary Radiation (a) | Assuming No Nuclear Collisions (b) | Primary Radiation Alone |
| 60,000 | 0.36 | 0.15-0.8 | 2.5(c) |
| 70,000 | 0.57 | | 6.2(c) |
| 75,000 | 0.68 | 0.4 -2.0 | 4  (d) |
| 80,000 | 0.80 | 0.6 -2.5 | 16  (c) |

(a) Data of Curtis.
(b) Data of Foelsche.
(c) Data of Fowler and Perkins for average dose over first hour.
(d) Data of Foelsche and Graul.

The results of a more detailed study of the sunspot numbers are presented in Figure 10-14. The sunspot number is an astronomical observation made from the surface of the earth which is corrected for the seeing at a particular observatory, and it is fairly well correlated with solar flares. These are polar

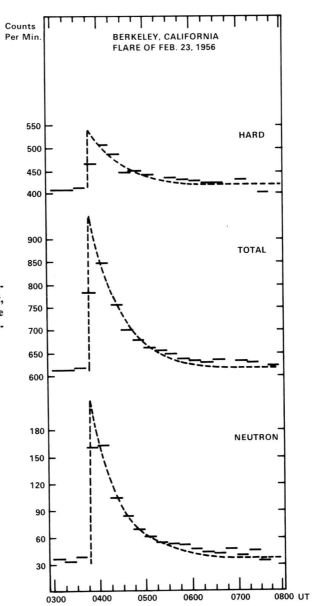

Fig. 10-16. The hard component, the neutron component, and the total flux during the large solar flare of 23 February 1956.

cap absorption (PCA) measurements made in the polar regions by simply measuring the incoming radio waves from the galaxy, which are absorbed when the ionosphere is distorted by a solar flare. It is a reasonably simple and straightforward measurement of a solar event. Many of them do occur during the height of the solar cycle. Listings of all the neutron radiation increases are also shown; they can be seen to occur largely during the peak of the solar

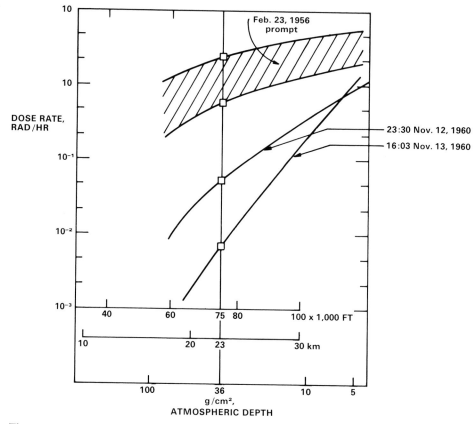

**Fig. 10-17. Dose rates within the atmosphere from solar flare particles (high magnetic latitudes).**

cycle. Hence it is probable that one can forget about solar protons for five or six years during every eleven-year solar cycle. Figure 10-15 shows how one can use detailed studies of the past two cycles to make reasonable predictions about the probable behavior of the next cycle. Calculated limits of sunspot number and solar flare events are shown.

A very famous plot of neutrons from the 1956 flare is shown in Figure 10-16 along with the rise in the hard component and the total radiation. The rise occurs in about half an hour; the time for a high-energy proton to arrive from the sun is about fifteen minutes; so the onset is not instantaneous. Actually the slope is such that it would allow action of some kind in maybe ten minutes if one had warning—for example, delaying a takeoff. The flare decreases in a matter of hours; some flare curves are flatter than this, but this is not too atypical. In Figure 10-17 flare readings have been converted to a dose rate in rads/hour. Compare this with the industrial tolerance level which allows 5 rads/year. The general population is allowed 0.5 rad/year. It is

**Table 17. Radiation Exposure to Occupants of SST in Relation to Altitudes in Feet (Values Are for Polar Latitudes).**

| Radiation | 30,000 | 40,000 | 50,000 | 60,000 | 70,000 | 80,000 |
|---|---|---|---|---|---|---|
| Galactic (maximum) | | | | | | |
| Charged particles and gamma rays | | | | | | |
| (mrads/hr) | 0.14 | 0.30 | 0.47 | 0.64 | 0.77 | 0.86 |
| (mrems/hr) | 0.24 | 0.44 | 0.67 | 0.95 | 1.2 | 1.6 |
| Neutrons | | | | | | |
| (mrads/hr) | 0.006 | 0.009 | 0.015 | 0.017 | 0.022 | 0.032 |
| (mrems/hr) | 0.048 | 0.075 | 0.116 | 0.134 | 0.172 | 0.254 |
| Total | | | | | | |
| (mrads/hr) | 0.15 | 0.31 | 0.49 | 0.66 | 0.79 | 0.89 |
| (mrems/hr) | 0.29 | 0.52 | 0.79 | 1.08 | 1.37 | 1.85 |
| Solar (avg) | | | | | | |
| (mrads/hr) | 0.001 | 0.004 | 0.014 | 0.045 | 0.15 | 0.45 |
| (mrems/hr) | 0.005 | 0.019 | 0.051 | 0.150 | 0.43 | 1.25 |
| Total (galactic plus solar) | | | | | | |
| (mrads/hr) | 0.15 | 0.31 | 0.50 | 0.70 | 0.81 | 1.34 |
| (mrems/hr) | 0.29 | 0.54 | 0.84 | 1.23 | 1.80 | 3.10 |

possible that a passenger flying at 70,000 ft during the flare might have received 0.2 rad which is about 40 per cent of the yearly dose. Perhaps one flare in five or ten years would be this size, so it seems likely that by use of the standards which allow 0.5 rad/year from manmade sources there is little danger. Another flare of lower intensity is also shown. The rules say that for the SST this is manmade radiation, although most of us look upon it as galactic and solar. Four other workers have estimated the 1956 flare dose rates in the region of 65,000 ft in altitude. Their results are shown in Table 16. Note that the more recent results are lower than earlier estimates and all are below 1 rad/hour. Additional estimates for polar latitudes are reported in Table 17 for several different components, again about 0.5 rad/hour in contrast with previous estimates which were somewhat higher. Probably we shall never know more about the 1956 flare than these rough calculations.

NEUTRONS

The same basic problem is encountered with the neutron component. Many neutrons have been measured but not at the latitudes and altitudes of the proposed SST routes. There have been measurements at sea level and various heights, including balloon altitudes which are above the SST by quite a bit. By

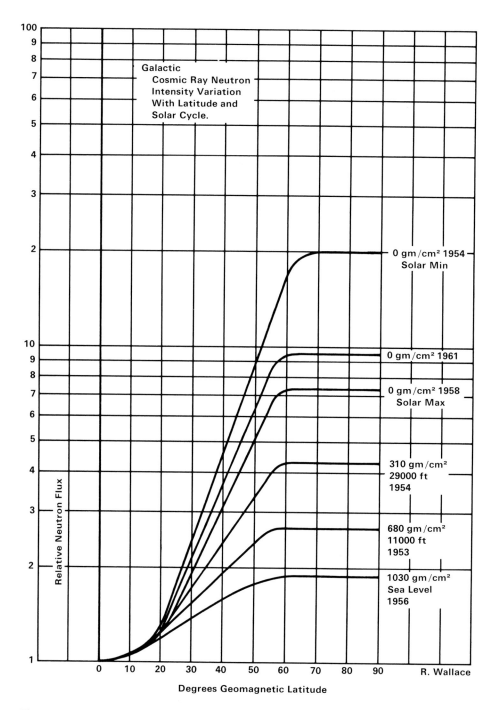

**Fig. 10-18. Galactic cosmic ray neutron intensity variation with latitude and solar cycle.**

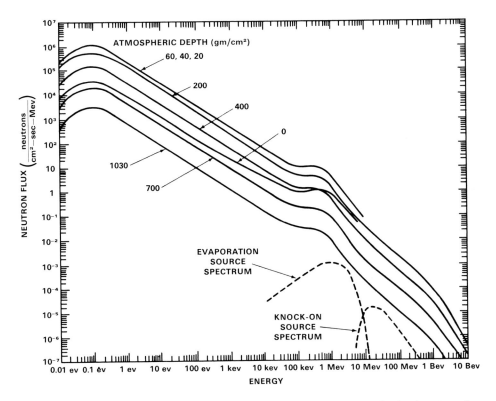

Fig. 10-19. The equilibrium neutron flux versus energy at different depths in the atmosphere for geomagnetic latitude $\lambda = 44°$. The energy spectra for 200 to 1030 gm/cm² are experimental values; for depths less than 200 gm/cm² the spectra are calculated. The shapes of the two neutron source spectra are also shown.

interpolating between balloon altitudes and aircraft altitudes, and assuming that the measurements made during the last solar cycle are reasonable, one gets the composite of neutron flux in Figure 10-18.

Figure 10-19 is a neutron spectrum which has been widely shown in the literature. One can see the evaporation peak, and the high-energy tail. Some measurements in the region above 50 MeV were made directly; some of this region is interpolated. There are some blanks and there are some measurements. Sea level is at 1030 gm cm⁻². The 200 gm cm⁻² is 40,000 ft. The SST is somewhere on the top curve. Multiplying this curve by the relative biological effectiveness of neutrons one can then get a dose distribution as a function of energy and an integrated dose. Such results are shown in Table 18. The dose lies principally between a fraction of, and 30 or 40, MeV.

If one makes measurements in the region of a few MeV one overestimates the neutron difficulties; while the neutrons become biologically more

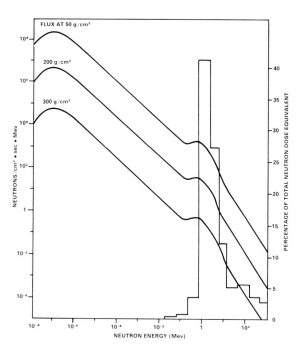

**Fig. 10-20.** Histogram of the neutron biological effect for galactic cosmic rays superimposed upon the neutron energy spectrum for three different atmospheric depths.

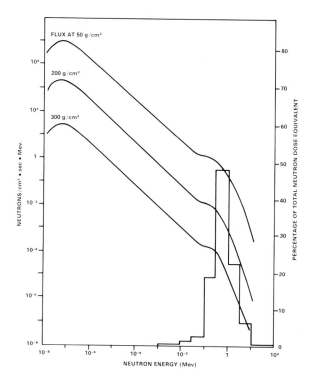

**Fig. 10-21.** Histogram of the neutron biological effect for solar cosmic rays superimposed upon the neutron energy spectrum for three different atmospheric depths.

**Table 18. Dosages from Galactic and Solar Cosmic Rays at Four Different Altitudes Based upon Relative Biological Effectiveness (RBE).**

(a) *40,000 feet*
    Galactic cosmic rays:   7 m-rads/day or 2.6 rads/year
                                       RBE $\cong 1.1$
    Solar cosmic rays:     $\sim$ 0.03 rads/year
                                         Worst flare per decade « 1 rad in one hour

(b) *60,000 feet*
    Galactic cosmic rays:   13 m-rads/day or 5 rads/year
                                         RBE $\lesssim 1.3$
    Solar cosmic rays:     0.4 rads/year
                                         Worst flare per decade $\sim$3 rads in one hour

(c) *80,000 feet*
    Galactic cosmic rays:   14 m-rads/day or 5.5 rads/year
                                         RBE $\lesssim 1.7$
    Solar cosmic rays:     5 rads/year
                                         Worst flare per decade 15 rads in one hour

(d) *100,000 feet*
    Galactic cosmic rays:   15 m-rads/day or 6 rads/year
                                         RBE $\lesssim 2.0$
    Solar cosmic rays:     36 rads/year
                                         Worst flare per decade $\sim$100 rads in one hour

effective at the high energies their number decreases rapidly. The neutrons become less effective at lower energy where their number is high, so there is a "focusing" of the neutron biological effect in this critical region, and thus neither a thermal detector nor a very high energy detector is useful. A detector for neutron energies between 0.1 MeV and 20 MeV is most effective in measuring neutron doses. Results for the galactic dose are shown in Figure 10-20. Note how little dose there is in the region above 50 MeV. Figure 10-21 shows the same calculation made for solar neutrons, and here it can be noted that above 50 MeV there's almost nothing, because the solar neutrons produced as secondaries from solar protons are so small in number above that energy. Therefore this dose curve has been shifted down a little bit in the case of the solar particles.

SUMMARY

From the standpoint of health physics we would like to know details of the radiation in the various categories for various altitudes, distributions such as charge particles, total neutrons, solar and galactic. At the altitude presently of interest, the total doses are in the neighborhood of millirads/hour on the average. A summary is given in Figure 10-22.

Many people fly frequently at 30,000–40,000 ft where the intensity is down by no more than a factor of 10 certainly. This raises the question which has been discussed over and over again—that if one worries at all about the dose at SST altitutdes one ought to worry almost equally about the dose at conventional jet aircraft altitudes, because it takes two or three times longer to fly the same distance in a subsonic jet and the dose is down only by a fàctor of 5 to 10. Perhaps all that has been said about the SST also applies to subsonic aircraft, so if one accepts it and realizes that these numbers are probably good to only a factor of 2 or 3, one should also accept the SST.

From the British-French data, the British have drawn several conclusions shown in Figure 10-23. (1) At ordinary jet aircraft altitudes (40,000 ft is a little high for a subsonic jet) the galactic cosmic rays produce about 2 rads/year, and the solar cosmic rays only 0.03. (2) At 60,000 ft this number is about 10 times higher, and the galactic cosmic rays are perhaps twice as

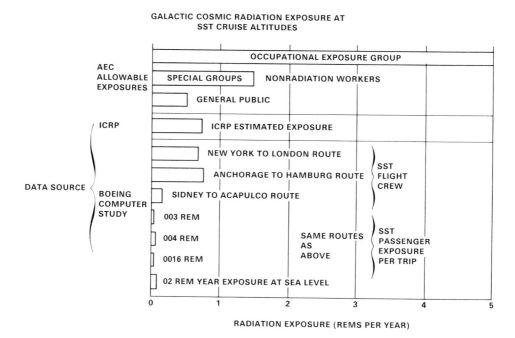

**Fig. 10-22. Galactic cosmic radiation exposure at SST cruise altitudes. Yearly estimated dose for flight crews flying polar routes 10 hr/wk (60,000 – 80,000-foot altitude) estimated by international commission on radiological protection. Boeing computer values are based on approximately 600 hours' flight time at 70,000 feet. It is assumed that crews fly the same routes.**

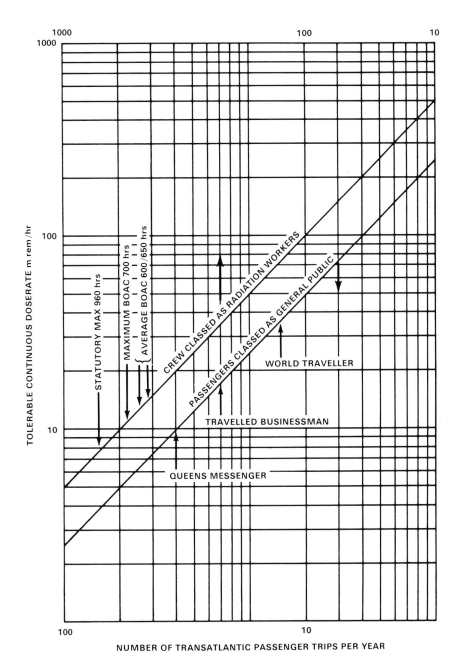

Fig. 10-23. British-French data and their proposed allowed dosages as a function of the number of trips per year at SST altitudes for various groups of personnel.

high. According to the British, the difference between ordinary jet aircraft and SST is about a factor of 2 for galactic cosmic rays. Galactic cosmic rays never increase very much so that their constant value prevents them from ever being a problem. There is a factor of 10 in the solar cosmic rays, but this means that if there are ten times as many passengers at ordinary jet aircraft altitudes there is the same genetic hazard. In any event it appears to be very small.

DISCUSSION

(Q) How does an SST trip compare with living in Denver, Colorado?

(A) Wallace: If you live in Denver I assume that you are there all the time. The dose in Denver is about 100 to 150 millirems/year due to radiation, so one flight during a solar flare roughly equals Denver. This problem has been studied by comparing Denver and San Francisco. They are cities of about the same size, but the radiation level differs by 50 or 100 millirems/year. It has been found that there are many more deaths from leukemia, presumably a radiation disease, in San Francisco than in Denver. Of course, if radiation were solely responsible Denver would have more. There is no evidence of any difficulties in Denver due to the radiation.

(Q) Perhaps it suggests there are other factors not yet known?

(A) Wallace: I think it proves that old people go to San Francisco to die of leukemia.

(Q) Is there any danger that during the year of high activity the crew might exceed the maximum dose before the year is over and as a result would be grounded?

(A) Wallace: No, I don't think so. Remember the crew is allowed 5 rems/year, and that is a pretty high dose. The number of hours they fly is rather limited. I have additional data on that if you're interested.

# Effect of Soil and Water on Cosmic Ray Neutrons JACOB KASTNER

*Abstract: Liquid scintillators, nuclear emulsions, and thermoluminescent crystals have been used to measure environmental neutron flux between thermal and 10 MeV energy below ground, at ground level, at air-water interface, on towers and at airborne altitudes of 20,000 ft. Preliminary results are compared with previous work and with calculations using a modified SNARG computer program.*

## INTRODUCTION

We have heard several papers concerning what goes on over our heads, now let us come down to earth. I would like to tell you our plans and progress in determining the contribution of naturally produced neutrons to man's environmental exposure. The fast neutron dose being received by man may be an important fraction of the total dose from the natural environment.

For many years, there was uncertainty in the literature with regard to the neutron component of the natural background. Over a decade ago, Marinelli, of our division at Argonne National Laboratory, proposed that this "numbers game" be resolved once and for all. (Incidentally, that was one of the early reasons for the existence of the Radiological Physics Division, namely, to supply precise and accurate numbers for politicians to fight with.)

In NBS Handbook 63 (1957) on Radiation and Protection, Marinelli provided an estimate of about 1.0 neutron cm$^{-2}$ min$^{-1}$ at sea level. This was based on extrapolations of upper atmosphere measurements such as have been discussed by the previous and learned speakers.

Since then, we perfected a liquid scintillator system that could give reasonably accurate values of the flux between from 10 to 15 MeV and 1 MeV as well as of the energy spectrum or distribution. It is important to know not only the total number but also the energy distribution from the standpoint of dose. The numbers we supplied are now considered to be the best in the world in this narrow region and proved to be lower than the earlier estimates by about a factor of 2.

Measurements by others verified this discrepancy and it was then that we started thinking of the interface questions. What happens to the neutron flux when one has soil nearby as well as air? What happens over water? Can one do anything to investigate the complete energy spectrum—right down to thermal energies? This question had been considered theoretically by Korff and Bethe in 1940.

This will be an account of the experimentation in progress as well as the thinking that led to the design of the experiment. There is much still to be

Fig. 11-1. The detection apparatus used for neutrons between 1 and 10 MeV. Proton recoils in liquid scintillator are selected electronically from the gamma background by the time duration of the light pulses.

done, and many of you have the equipment and facilities to carry out other investigations along similar lines. I hope this talk will stimulate some of you to work along parallel paths, after you see what we hope to accomplish.

NEUTRONS FROM 1 TO 10 MEV

I shall separate the work into three energy regions, because no single detector can actually cover the complete energy region. These divisions are 10-1 MeV, 1 MeV to 1 KeV and below 1 KeV. I shall talk mostly about 1-10 MeV and leave the rest to my colleague, Dr. Ray Gold.

The apparatus used in the region of 1-10 MeV is shown in Figure 11-1. The liquid cell consists of 1600 cc of N.E. 213 scintillator with a conical light pipe to a 2-in. photomultiplier. The base contains an electronic circuit which discriminates between gamma rays and neutrons on the basis of the duration of the light pulse. When a neutron collides with a hydrogen atom

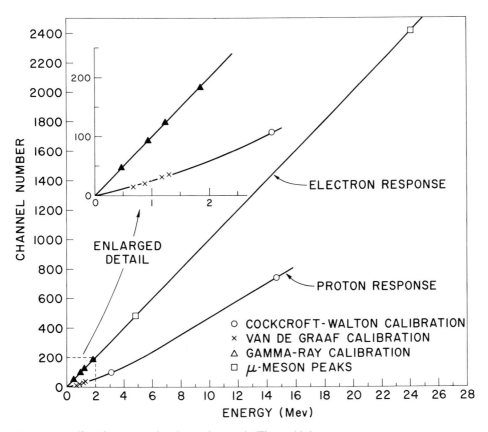

Fig. 11-2. Calibration curves for the equipment in Figure 11-1.

of the scintillator liquid, the proton recoil at these energies is slow and the specific ionization is high. The light from such an event takes a long time to decay. Conversely, for relativistic ionizing particles, the light pulse is of short duration. The circuit is a commercial version of a development in 1961 of Debwick and Sherr and works very well. One can, with confidence, discriminate for 1 proton in a field of 1000 gammas. The pulses are analyzed on a 400-channel T.M.C. analyzer and the distribution either typed out or punched out on tape. The system has been calibrated as shown in Figure 11-2. The point at 25 MeV is a natural calibration due to mesons which are relativistic and pass through the cell vertically. They, therefore, deposit a known energy of about 2 MeV $gm^{-1}$ $cm^{-2}$ in a single traverse. The channel number is really a measure of light amplitude.

We carried out measurements in the laboratory over long periods to get good statistics. Occasionally, we were bothered by manmade neutrons from our friends in Reactor Physics or Neutron Physics.

The numbers appear in Table 19 and were lower than the early measurements of Hess, Patterson, and Wallace but were in fair agreement with a sea-level value extrapolated from Hayme's balloon measurements above 10,000 ft. Hayme's number for the total flux is in reasonable agreement with our value 0.4. Data from other workers are included for comparison, and are in agreement within a factor of 3. The problem with such extrapolation and with any measurements on the ground, of course, is the one raised by everyone in this business, namely, what is the effect of the earth on neutron flux?

COMPUTER CALCULATIONS

This last year we were host to a visiting scientist from Israel and he and I started to examine this problem in earnest. One advantage of a place like Argonne is that within easy reach there are experts at everything. If we approach them properly and employ an appropriate soft-sell, they may even do a little work for us, especially if we can interest them in the problem itself. Dr. Kelber in Reactor Physics, for example, suggested we analyze our problem using their computer program for reactor studies. This is called or coded SNARG (SN calculation at Argonne) and is a complex transport equation calculation. After tailoring the input to suit our tenuous gas conditions rather than the solids and liquids, we did indeed obtain a description of what we could expect over soil and water for a number of neutron energy groups.

Figure 11-3 shows the results for neutrons between 2 and 4 MeV. The

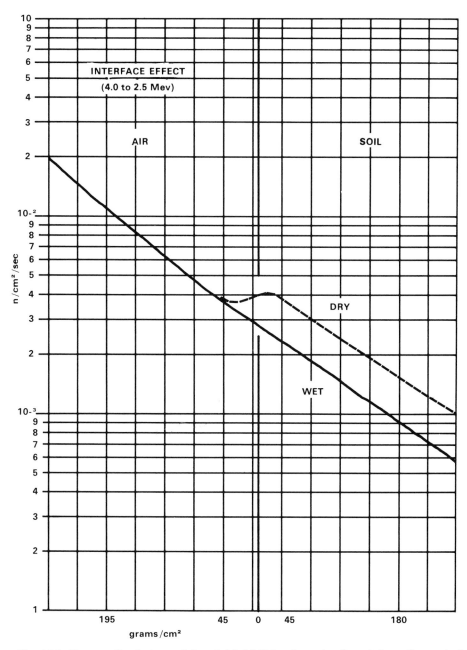

**Fig. 11-3. Neutron flux between 2.5 and 4.0 MeV in air, wet soil, and dry soil as calculated by use of the modified SNARG program.**

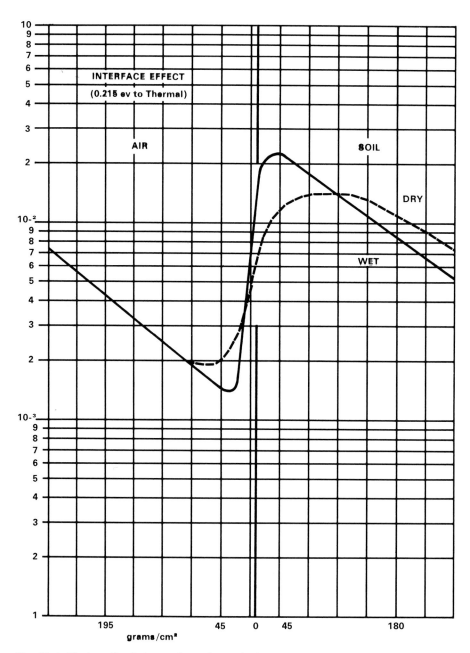

**Fig. 11-4. Neutron flux between thermal and 0.215 eV in air, wet soil, and dry soil as calculated by use of the modified SNARG program.**

**Table 19. Comparison of Environmental Neutron Flux Measurements of Various Observers.**

| Observer | Cosmic-Ray Neutron Flux $n/cm^2/sec$ | | Relative to Hess |
|---|---|---|---|
| | 1-10 MeV | Total | |
| Hess et al. | 0.01 | 0.041 | 1 |
| Edge | | 0.013 | 0.31 |
| Kent | | 0.0087 | 0.21 |
| Boella et al. | | 0.0065 | 0.16 |
| Yamashita | 0.0017 | 0.0073 | 0.18 |
| Haymes | 0.005 | | 0.5 |
| This experiment | 0.0038 | | 0.38 |

interface is the vertical line in the center. Since the abscissa is in gm cm$^{-2}$ the curve begins at the left at 10,000–15,000 ft above the earth. If the soil is wet the flux is little different from an extrapolation of the air curve. For dry soil there is a 30 to 50 per cent increase at the earth's surface. (We were unable to calculate results for pure water because the SNARG program failed for this case.)

The results for the thermal to very low energy group are shown in Figure 11-4. For neutrons of energy below 0.25 eV the effect is enormously magnified. We obtained factors of the order of 20 near the earth's surface. The reason is that the nitrogen in the air has a large cross-section for low-energy neutrons so that the curve in air for this energy is far below the air curve in Figure 11-3. After the remaining neutrons enter the earth there is a large thermal contribution boiling back out of the earth.

THE AIR-WATER INTERFACE

It was immediately obvious that measurements over water would help to check some of these calculations. So we enlisted the cooperation of the City of Chicago Water Department, and they let us use a room in one of their Lake Michigan intake cribs. Figure 11-5 shows part of one of the cribs, about 4 miles from the shore east of Chicago. For fifty years several of these cribs were used as intakes for the city water supply, but they are now being phased out. Like castles surrounded by moats they accommodated as many as thirty-five workers in baronial splendor in a manner befitting an officer's mess. Now there are only two caretakers on duty for a week at a time. The cribs are free from manmade radiation and electrical interference from power lines. The city furnished both telephones and power. Not only that, but they take

Fig. 11-5. One of the Chicago Water Department intake cribs, as seen from the deck of an ice-breaker ship, provided an isolated off-shore neutron detection station for air-water interface measurements.

Fig. 11-6. Doctors Fiege and Kastner operate the equipment illustrated in Figure 11-1 at the station shown in Figure 11-5.

**Fig. 11-7. Oltman attaches cans containing emulsions and thermoluminescent crystals to the superstructure of the water intake crib.**

us out on an ice breaker whenever we wish and even let us train one of their caretakers to punch out the data every week. We set up our apparatus shown in Figure 11-6. The one with hair is Dr. Yehuda Fiege of Israel AEC. I'm fiddling with the detector for the photographer's benefit.

At the same time we set out some thermal neutron detectors in the form of $Li^6$ and $B^{10}$ loaded emulsions and $Li^6F$ thermoluminescent crystals. In Figure 11-7 you see my colleague, Billie Oltman, fixing a can containing these detectors to the railing of a bridge at the crib, 30 ft above deep water. Similar cans were placed at Argonne sites, underground, at ground level, and on the top of a 30-m ranger tower.

In order to confirm the thermal neutron variation above soil and water we devised another experiment in the region of the interface. We dug a 5-in.-diameter hole at a 45° angle and 2 m into the earth. Using a photomultiplier with a thin glass scintillator loaded with Li-6 or Li-7, we measured the thermal neutron flux at 10-cm intervals in an effort to reproduce the profile predicted by the computer calculations. Our electronic equipment worked very well, but the counting rate was low and the statistics are rather poor.

To complete the picture we needed more data points in order to check the fit with our theoretical expectations.


AIR-BORNE MEASUREMENTS

We decided to fly our 5 x 5 scintillator at 20,000 ft where we could be well away from the earth or water (a half-atmosphere away). We negotiated with the hurricane spotters of ESSA at Miami to let us fly on their DC-6 for a few hours. They have a 60-cycle generator, and all the meteorological data are continuously and automatically recorded. This flight took place two weeks ago today and we were able to obtain in four hours data which normally require eleven days to accumulate on the ground or over a month on the lake.

The data are as follows for 1-10 MeV neutrons:

|  | Flux | Dose $\mu$rad/hr |
|---|---|---|
| On land: | 0.004 n/cm$^2$/sec | 0.032 |
| On lake: | 0.016 n/cm$^2$/sec | 0.013 |
| 20,000 ft: | 0.25 n/cm$^2$/sec | 2.0 |

These are preliminary results and are admittedly still imprecise. However, one fact is clear—the 100,000 lb of airplane, mostly aluminum, constitutes an environment almost identical to the ground. This is because the primary cosmic ray attenuation length in air is somewhere near 140 gm/cm$^2$; this fits almost exactly with the factor of 65 increase in intensity over ground. It doesn't fit at all with the factor of 150 over water.

Finally, we obtained a pretty good fit for the thermal neutron profile in the ground, so good, in fact, that we think we can detect the effect of soil moisture on the profile. We plan to try and use this as a passive soil-moisture meter.

I regret to report that the cans of emulsions and thermoluminescent crystals were not sealed well enough to weather well over the long exposure period and we lost our first set of data. These experiments are being continued.

# Cosmic Ray Neutrons and Carbon-14 Production

## RAYMOND GOLD

*Abstract: The natural production of radiocarbon by cosmic ray neutrons has been observed experimentally. Proportional counters, filled with two atmospheres of nitrogen, have been employed to detect the exoergic $N^{14}(n,p)C^{14}$ reaction. Electronic pulse-shape discrimination is utilized to reject meson and photon-induced events. Proton-recoil measurements of the sea-level spectrum of cosmic ray neutrons in the energy region 0.05-2.0 MeV have also been carried out using this detection system with three atmosphere fillings of hydrogen and methane. In general, the neutron flux per unit energy decreases monotonically with increasing energy. The resulting geophysical data are:*

1. *$C^{14}$ production rate:*      $(1.89 \pm 0.08) \times 10^{-7}$ /sec/cm$^3$
2. *Slow neutron density:*     $(1.19 \pm 0.06) \times 10^{-8}$ n/cm$^3$
3. *Integral fast neutron flux (0.05-2.0 MeV):*     $2.3 \times 10^{-3}$ n/sec/cm$^2$

*for the air-land interface at the Argonne National Laboratory site (53° N. geomagnetic latitude).*

INTRODUCTION

I shall report on some cosmic ray studies and measurements which have been performed at Argonne National Laboratory during the past four years with the support of the Radiological Physics Division and the Reactor Physics Division.

Professor Korff has delineated the problem of intermediate neutron energies below 1 MeV related to the lack of definition of the spectrum of neutrons in this energy region. We are actively engaged in developing instruments to measure neutrons in this region. The experimental technique I shall describe has been developed in collaboration with personnel in the Reactor Physics Division, principally Ed Bennett.

INTERMEDIATE NEUTRONS AT SEA LEVEL

Gas-filled proportional counters employed the classical hydrogen recoil process in hydrogen gas and methane. The problem of discrimination against background events was encountered and was solved by pulse-shape discrimination to reject photon-induced and meson-induced events. This method depends upon two facts: (1) The total ionization is proportional to the total energy, (2) The rise-time and time-dependent slope of the pulse depends upon the specific ionization. The hydrogen recoil from a neutron is very localized, if the gas pressure is properly chosen, and has a very fast rise-time. In contrast with this, the high-energy events create long tracks and the gammas produce primarily wall electrons, both of which cause an extended distribution of ionization with slower rise-times.

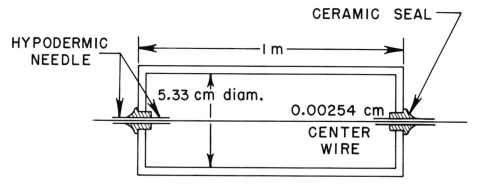

Fig. 12-1. Geometric characteristics of the 4π hydrogen-recoil proportional counter. Figures 12-1 through 12-8 reprinted, with permission, from Gold, *The Physical Review*, vol. 165, no. 5 (Jan. 25, 1968), pp. 1406-1411.

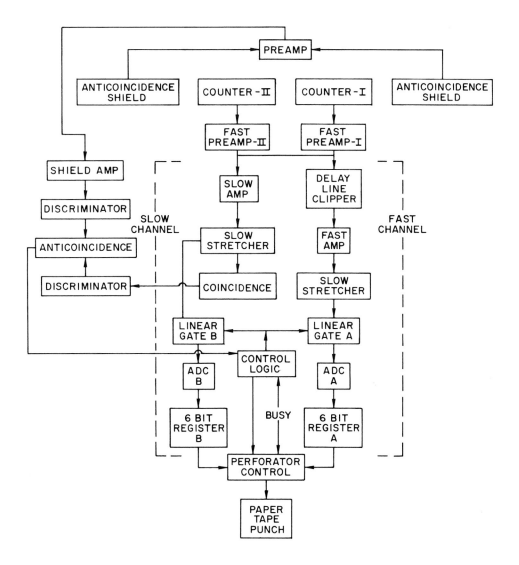

**Fig. 12-2. Logical block diagram of the cosmic ray neutron two-parameter analyzer.**

The energy region of interest determines the anode voltage chosen for operation. We used 3000–4000 V. The counters are 5.72 cm in diameter with 0.194 cm stainless steel wall. The active length of each counter was approximately 1 m. A 0.00254-cm-diameter stainless steel wire served as the anode. The anode wire was centered in the cathode by means of hypodermic needles which were, in turn, positioned in ceramic seals at each end of the counter body. The geometric characteristics of these $4\pi$-proportional counters are shown in Figure 12-1.

Preliminary sea-level measurements at the air-land interface which emphasize the serious background problem are:

| Energy | Background/Signal |
|---|---|
| 60 KeV to 240 KeV | 500:1 |
| 240 KeV to 480 KeV | 100:1 |
| 300 KeV to 1200 KeV | 4:1 |

One is faced with an increasing background problem at lower neutron energy.

The instrumentation used is shown in the logical block diagram of Figure 12-2. The slow channel gives the energy information and the fast channel gives the rise-time information. It is clear that two-dimensional data analysis is required. Such multiparameter nuclear instrumentation is expensive principally owing to the need for the memory unit, usually called a pulse-height computer. At our low counting rates we could eliminate this costly component by use of punch-paper-tape recording. In fact, many important measurements can be made at lower event rates, where one can use a paper-tape system and avoid the expense of a large computer-type memory.

The energy calibration of the equipment is shown in Figure 12-3. The reaction $N^{14}(n,p)C^{14}$ is exoergic at 615 KeV. A small amount of nitrogen was introduced into the proportional counter gas for counter calibration and evaluation of counter operation.

There were three experiments in this series at Argonne National Laboratory during 1966. The first two experiments had counter fillings of pure methane to which small amounts of calibration nitrogen were added. Various high voltages were used to cover different energy regions. The third experiment used hydrogen gas with a small admixture of methane for quenching and a bit of nitrogen for calibration; this carried the low-energy region down to 40 KeV. The limit was set by the event-rate which could be handled by the paper tape memory unit. The general characteristics of these three experiments are as follows:

**Table 20. General Characteristics for Experiments 1-3.**

| Experiment no. | Partial Pressure (psi) | | | Energy Region (MeV) | Live-time Duration (sec) |
|---|---|---|---|---|---|
| | $CH_4$ | $H_2$ | $N_2$ | | |
| 1 ............. | 45.0 | ... | 0.20 | 0.500-3.10 | $2.8968 \times 10^6$ |
| 2 ............. | 45.0 | ... | 0.20 | 0.120-0.690 | $3.5307 \times 10^6$ |
| 3 ............. | 4.0 | 41.0 | 0.75 | 0.040-0.144 | $2.8680 \times 10^6$ |

Figures 12-4, 12-5, and 12-6 show the proton-recoil data obtained in experiments 1, 2, and 3, respectively.

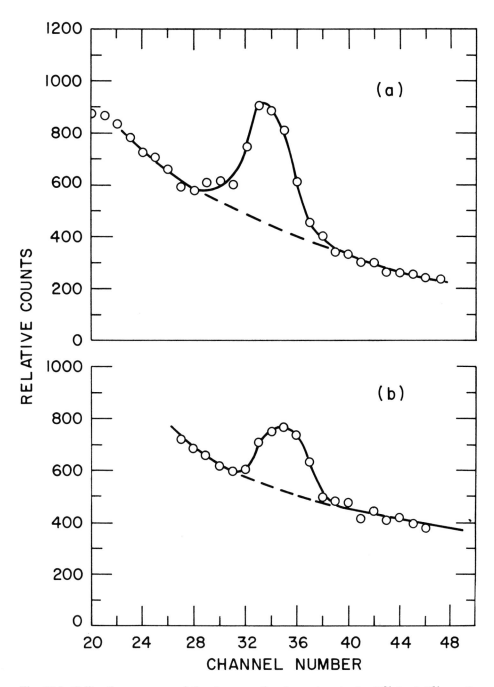

Fig. 12-3. Calibration response of the $4\pi$-proportional counters to the $N^{14}$ (n,p) $C^{14}$ reaction for (a) counter I, experiment 3, 3200 V; and (b) counter II, experiment 1, 2800 V.

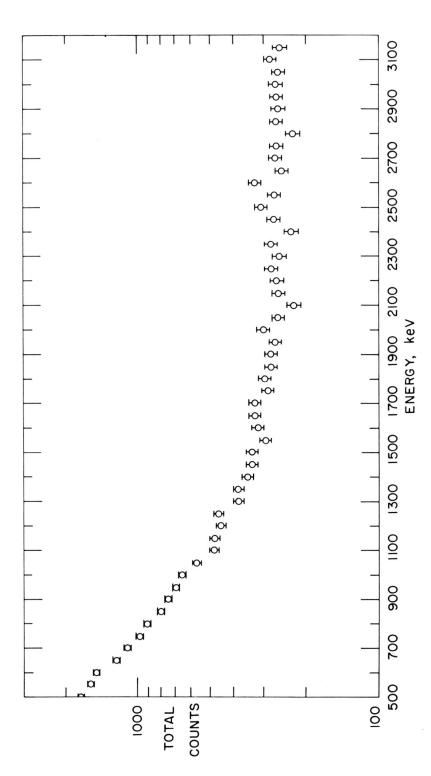

Fig. 12-4. The proton recoil data between 0.500 and 3.10 MeV.

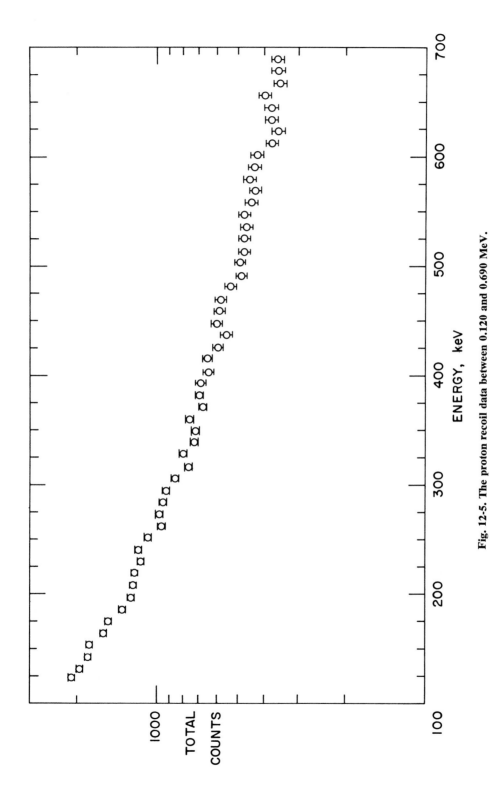

**Fig. 12-5. The proton recoil data between 0.120 and 0.690 MeV.**

ENERGY, keV

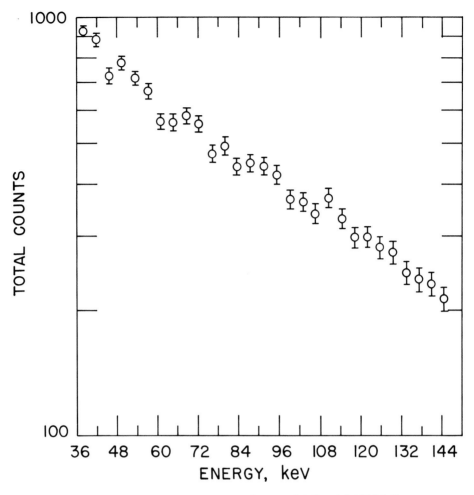

**Fig. 12-6. The proton recoil data between 0.040 and 0.144 MeV.**

For our specific experiments Figure 12-7 contains an example of the resulting cosmic ray count rate distribution. The abscissa is A/B which is proportional to specific ionization where A is the rise-channel and B is the fixed energy channel of each respective measurement. One sees two overlapping distributions. The sharp distribution at very low values of specific ionization are the electrons, mesons, and other extended track length events. The solid black points are $\gamma$-rays which are typical of long track length events in the counter. They are subtracted out so that proton recoils can be analyzed.

In Figure 12-8 is shown the composite neutron spectrum at the air-land interface from 0.050 to 2.50 MeV. The larger error flags toward the right is an indication of the inability of the detection system to sense any flux above

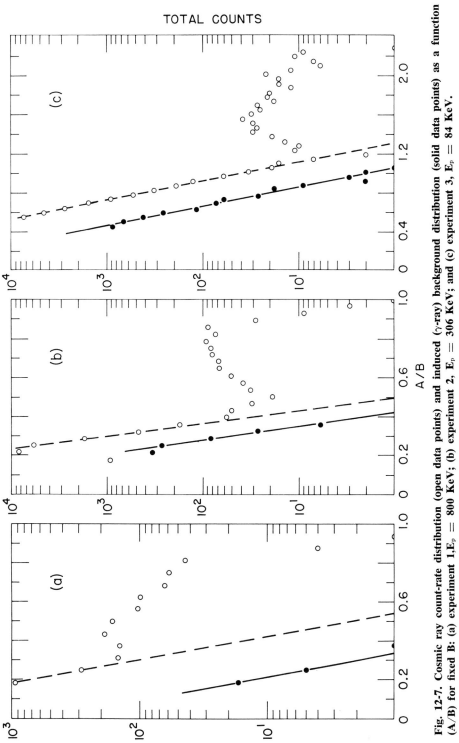

**Fig. 12-7.** Cosmic ray count-rate distribution (open data points) and induced ($\gamma$-ray) background distribution (solid data points) as a function (A/B) for fixed B: (a) experiment 1, $E_p = 800$ KeV; (b) experiment 2, $E_p = 306$ KeV; and (c) experiment 3, $E_p = 84$ KeV.

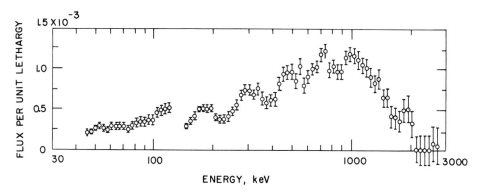

**Fig. 12-8. Composite neutron spectrum in the energy region 0.050-2.50 MeV determined from experiments 1, 2, and 3. The customary definition of lethargy has been used, $\mu = \ln(E_0/E)$, where $E_0$ is an arbitrary reference energy.**

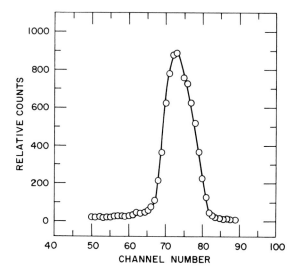

**Fig. 12-9. Calibration response of counter I due to $N^{14}$ (n,p) $C^{14}$ reaction in a 30-psi nitrogen filling with 3600 V on the anode.**

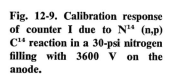
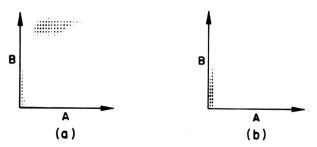

**Fig. 12-10. Cathode ray oscilloscope displays of the two-dimensional (A x B) pulse-height response for: (a) $N^{14}$ (n,p) $C^{14}$ reaction, and (b) $CO^{60}$ $\gamma$-source.**

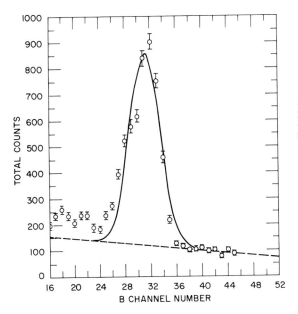

Fig. 12-11. Experimental pulse-height distribution obtained from mid-November to mid-December 1966 at the Argonne site. The smooth curve is a Gaussian fit to the experimental data above the background continuum.

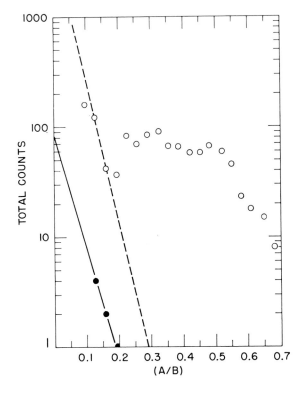

Fig. 12-12. The observed count-rate distribution (open circles) and the induced (γ-ray) background distribution (solid circles) as a function of (A/B) for B = 31.

about 2 MeV. There is some structure in the spectrum, notably the pronounced dip near 400 KeV which corresponds to scattering resonances in N and O both in the atmosphere and in the soil. It is typical of a scattering resonance, in which the neutrons skip across the resonance and build up on the lower side. The lower resonances may be due to Si.

CARBON-14

As an outgrowth of our experimental technique, it is possible to make direct measurements of the $C^{14}$ production rates at sea level. This is the most abundant isotope produced in the atmosphere by cosmic rays and is part of our natural environment. One of our proportional counters was filled with pure nitrogen. Figure 12-9 shows the response of the counter to a calibrated hand neutron source. One sees here the proton in the (n,p) reaction which carries off about 585 KeV plus a very small recoil of the $C^{14}$. It is very neat and clean to obtain such a calibration, but it is quite another problem to look for the reaction in the natural environment.

The excellent capabilities of the electronic pulse-shape discrimination system are shown in Figure 12-10. Part (a) is the cathode ray oscilloscope display from a one-minute time exposure with the neutron source. It exhibits the wide separation of the $\gamma$-ray events (close to the B-axis) and the proton events (well away from the B-axis). Part (b) is a three-minute exposure to only a $Co^{60}$ $\gamma$-ray source.

Our two-atmosphere counters were operated for over one month to give the peak shown in Figure 12-11. The B-parameter was set to cause the peak to occur in pulse height channel B $= 32$. The measured background has been subtracted out to give the Gaussian fit shown by the solid curve. It is assumed that the background is from fast neutron knock-ons directly on the N in the counter. The background correction for each B-channel data point was carried out by studying the count rate distribution as a function of A/B, similar to the method shown in Figure 12-7. An example of just one channel, B $= 31$, is illustrated in Figure 12-12.

We were able to determine that the absolute $C^{14}$ production rate at the earth's surface is $(1.89 \pm 0.08)$ x $10^{-7}$ sec$^{-1}$ cm$^{-3}$. We are presently engaged in extending these measurements to other environments.

DISCUSSION

(Q)  What is the lethargy in Figure 12-8?

(A)  Gold: A common scale used by reactor people for many years. It is essentially log $(E_0/E)$ where $E_0$ is arbitrary. If we divide the data in Figure 12-8 by energy, we obtain flux per unit energy. One convenient characteristic of a lethargy scale is that of a weakly absorbing medium with no sources present. One finds a $1/E$ spectrum for a central solution to the Boltzmann problem. A $1/E$ spectrum on a lethargy scale is a constant.

(Q) Are your $C^{14}$ results for dry air or nitrogen?

(A) Gold: We obtain in the counters the production rate per N atom, which we scale to one atmosphere at sea level.

(Q) At what latitude and longitude were your measurements made?

(A) Gold: Argonne is $42°$ N. and $88°$ W.

(Q) Does this $C^{14}$ result give you a way at getting at the sunspot numbers in ancient time?

(A) Gold: There are problems of the total $C^{14}$ inventory. To date there are many measurements of $C^{14}$ in the atmosphere and in the past the production rate has been inferred from neutron measurements. Our measurements will improve the accuracy of the world-wide production rate, and therefore the balance in the equation may be based upon a firmer foundation, but I don't see the relevance to sunspots.

(Q) In monitoring $C^{14}$ for military purposes it is not known whether the fluctuations are due to natural causes or bombs. This work of Gold's establishes an accurate base line for natural production.

# V. Comets and Dust

# Interstellar Dust
## ADOLF N. WITT

*Abstract: Photoelectric measurements using three different filters are used to determine the intensity and distribution of the diffuse galactic radiation (DGR) which originates in the interaction of starlight with interstellar dust. The results are compared with radiative transfer calculations using the albedo γ and scattering asymmetry factor g for four different models of interstellar particles. The Platt model gives the best fit.*

INTRODUCTION

Figure 13-1 shows a field in our Milky Way. The dark patches superimposed upon the star background are clouds of dust particles which extinguish the light of the stars behind them. It is highly unlikely that these are actually holes in the star distribution, and the explanation that they are dust clouds is the only reasonable one. Furthermore, studies of the light from the stars which are apparently behind the clouds show it has undergone significant changes which we shall discuss in more detail.

In interstellar space starlight is interacting with dust particles. This interaction, which may be absorption or scattering or both, changes the quantity and quality of the direct starlight and converts at least part of the light incident upon the dust into a diffuse radiation field.

**Fig. 13-1. A star field in our Milky Way. The dark patches superimposed upon the star background are clouds of interstellar dust.**

Studies of the changes of the direct starlight—the interstellar extinction—indicate that the magnitude of this effect is approximately inversely proportional to the wave length throughout the observed range from about 1000 Å to 20,000 Å. This fact combined with arguments based on mass limitations excludes Rayleigh scattering (which has a $\lambda^{-4}$ dependence), electron scattering, or blocking by large particles as possible causes for the interstellar extinction.

Models capable of explaining the observed phenomena are based on Mie scattering—classical electromagnetic scattering by spherical isotropic particles with dimensions comparable to the wave length of the incident light. A variety of substances—iron, ice, dirty ice, graphite, and graphite covered with ice—each in suitably chosen size distributions have been suggested in the past. Usually, a change in material was in order whenever observations of the interstellar extinction law became available for a wave length range larger than previously observed. This feature of the Mie scattering models for the interstellar extinction is a most disturbing one.

**Fig. 13-2. Our galaxy would appear very similar to this figure if viewed edge-on from another galaxy.**

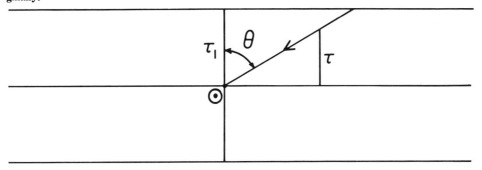

**Fig. 13-3. The slab model of our galaxy as used to analyze the radiative transfer. The sun is represented by O.**

A totally different explanation for the effects caused by interstellar dust was suggested by Platt (1956). He proposed that particles of less than 20-Å diameter have quantum-mechanical optical and statistical properties which might well explain the observed interstellar reddening.

Although the different models predict a very similar wave length dependence of extinction, they vary considerably in their scattering properties, especially in their albedo and in the shape of their phase function.

These properties can be determined observationally from a study of the diffuse galactic radiation (DGR) originating in the interaction of starlight with interstellar dust through the galaxy, as first attempted by Henyey and Greenstein (1941).

## THEORY

If we could view our galaxy from a great distance away it would appear very similar to the edge view of the galaxy shown in Figure 13-2. Our earth is about two-thirds of the way out from the center and approximately in the median plane. We shall idealize this by a slab model of Figure 13-3 in order to analyze the radiative transfer in our galaxy. It has finite optical thickness in the $\tau_1$ direction and is essentially infinite in the plane of the slab. The theory shows that the transfer equation is

$$\cos \theta \, \frac{\mathrm{dI}}{\mathrm{d}\tau} = \mathrm{I} - \int \mathrm{I}\phi \mathrm{d}\omega - \mathrm{a} \qquad \text{Equation 1}$$

where I is the intensity of starlight and DGR, $\theta$ and $\tau$ are defined in the figure, $\phi$ is the phase function of the scatterer and hence depends upon the material, and a = (emission coefficient/volume/solid angle) ÷ extinction coefficient/ unit length. The phase function depends upon two parameters, $\gamma$ the albedo and g the asymmetry factor so that it becomes $\phi = \phi(\gamma, g)$. The limits on $\gamma$ are zero for complete absorption to one for complete scattering; g can have values $-1 \leq g \leq +1$ where $-1$ corresponds to complete back scattering for large particles and $+1$ to complete forward scattering.

This equation can be solved analytically if a is sufficiently constant throughout the galaxy.

## OBSERVATIONS

Let us now turn to the observations. The observed intensity is the sum of seven separate contributions

$$\mathrm{I_{obs}} = \mathrm{I_o} + \mathrm{I_1} + \mathrm{I_{zod}} + \mathrm{I_{AG}} + \mathrm{I_{at}} + \mathrm{I_{art}} + \mathrm{I_{aur}}$$

where
- $\mathrm{I_o}$ = intensity of starlight
- $\mathrm{I_1}$ = intensity of diffuse galactic light
- $\mathrm{I_{zod}}$ = intensity of zodiacal light
- $\mathrm{I_{AG}}$ = intensity of airglow
- $\mathrm{I_{at}}$ = intensity of light scattered in atmosphere
- $\mathrm{I_{art}}$ = intensity of scattered artificial light
- $\mathrm{I_{aur}}$ = intensity of auroral light.               Equation 2

The last two can be made negligible by proper choice of observatory location and time at which observations are made. With careful choice of observing time intervals and regions of the sky $\mathrm{I_{AG}}$ and $\mathrm{I_{at}}$ can be kept constant. Recent

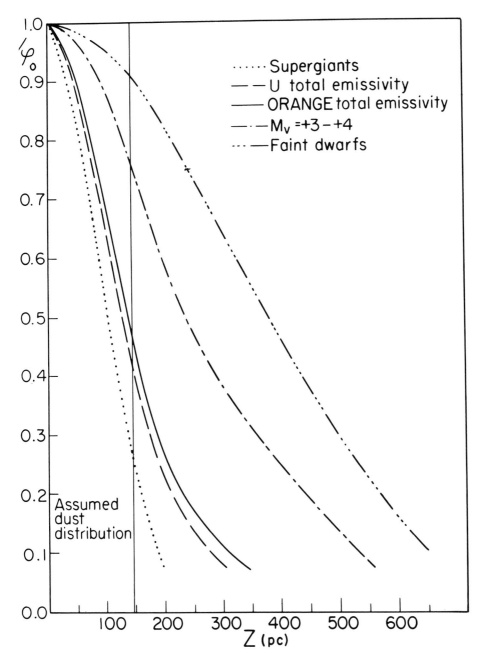

**Fig. 13-4. Relative distribution of interstellar dust and different star types with height above the galactic plane as used in the galaxy model.**

measurements of $I_{zod}$ can be used for the correction term or one can choose regions far from the zodiac. $I_o$ is determined from available star counts. Thus from $I_{obs}$ one can obain $I_1$.

Figure 13-4 shows the normalized distributions of interstellar dust assuming a constant value of a and various types of stars having a wide range of luminosity. The total emissivity per unit volume is determined more strongly by the more luminous stars.

The star field is so dense in the Milky Way (see Figure 13-5) that it is necessary to limit the fields to diameters of minutes arc in order to avoid contributions from even the faintest stars.

Photoelectric measurements in three colors were made of the diffuse galactic light with the 36-in. and 82-in. telescopes of the McDonald Observatory. Predetermined fields with diameters of 67 sec of arc were measured in the general directions of $1^{II} = 170°$ and $70°$ with galactic latitudes ranging from $b^{II} = -35°$ to $+35°$ in Tau-Aur and $b^{II} = -30°$ to $+30°$ in Cygnus. The fields were chosen under the restraint of constancy of air mass during all observations of a given night. The presence of a faint BD-star within a distance of 4 minutes of arc as reference for offsetting was required and, lastly, no star brighter than $V = 20$ should be present in the diaphragm, the latter being established with the help of the Palomar Sky Survey prints.

The band-passes used were those of the U and B filters of the UBV system with a 1P21 multiplier as detector and that of a red interference filter centered at 6100 Å with 275 Å half width in combination with an ITT 130-G multiplier (pass-band 0 for ORANGE). With this arrangement strong night sky emissions were thus generally avoided. These are in Table 21 which shows the intensity of different light sources near the galactic equator in Cygnus for the three wave length band-passes used in the photoelectric measurements.

The variations of the general sky brightness were monitored throughout the nights and standard stars were observed for calibration and extinction determination. Under these conditions several scans across the Milky Way in both Cygnus and Tau-Aur were measured repeatedly in November-December 1965 (U,B) and June-July 1966 (U,B,O).

RESULTS

After corrections for sky brightness variations, extinction and scattering in the atmosphere, zodiacal light, and in a few cases for faint background stars (unavoidable in Cygnus) were applied to the measured intensities, it was assumed that the amount of diffuse galactic light was small at $|b^{II}| > 27°$, the radiation

**Table 21. Intensity of Different Light Sources Near the Galactic Equator in Cygnus [a].**

| Source | U | B | ORANGE |
|---|---|---|---|
| Airglow and zodiacal light | 140 | 80 | 225 |
| DGR | 45 | 40 | 90 |
| Faint stars B⩾17 | 5 | 9 | 40 |
| Faint stars B⩾20 | 1 | 2 | 6 |
| Total star background | 160 | 170 | 500 |

(a) The units are numbers of AOV stars of V = 10.0 per square degree in the respective band-pass.

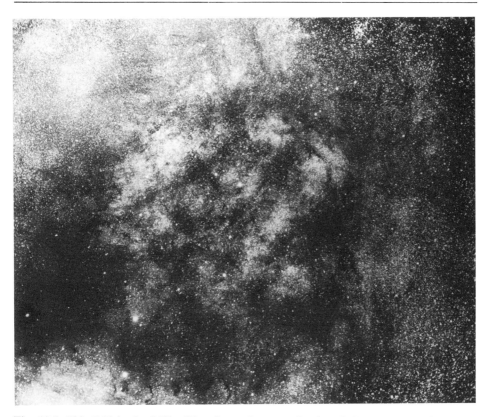

**Fig. 13-5. This field in the Milky Way shows the great density of the star population which required that the field of observation be limited to a diameter of minutes of arc.**

detected there being due primarily to air glow and zodiacal light. The resulting intensities above this level are taken to be lower limits to the diffuse galactic radiation. Results of these measurements are shown in Figure 13-6.

In Cygnus a well-defined maximum of intensity is found at $b^{II} = +5°$, the intensities being $n(U) = 45$, $n(B) = 40$, $n(O) = 100$ (the uncertainties are estimated to be within 10 per cent) in units of number of stars of U = 10.0, B = 10.0, and O = 10.0 per square degree, respectively. The somewhat less well-determined maximum values in the Tau-Aur region are n(U) = 40 and n(B) = 35. The integrated star brightness is assumed to be in the ratio B:U:O = 1.00:0.94:2.94 at $b^{II}$ = O°, a result which follows from the properties of a galaxy model having a luminosity function and a star distribution as given by Schmidt (1959). This model was used also to compute the star distributions and the distribution of interstellar dust as functions of distance from the galactic equational plane as shown in Figure 13-4.

ANALYSIS

Several methods for analyzing the data and comparing the results with predictions from Henyey and Greenstein's (1941) model for the radiative transfer in the galaxy have been studied. The final approach was that of a detailed comparison of the observed intensity distributions with theoretical distribution curves of the diffuse galactic light. These were computed for the Cygnus and the Tau-Aur regions for the three filter band-passes, assuming scattering characteristics of the different models for interstellar dust particles, which have been considered in the past. The results are shown in Figure 13-7. These scattering properties, specifically the albedos and the asymmetries of the phase function, were derived from Mie's theory for spherical homogeneous and com-

**Table 22. Albedo and Asymmetry Factor of the Scattering Phase Function for Different Particle Models.**

| Model | λ | γ | g |
|---|---|---|---|
| Dirty ice, Oort-v.d. Hulst distr. n = 1.33-0.05i $r_H = 0.125\mu$ | 3597 4348 6100 | .73 .69 .60 | .76 .71 .58 |
| Graphite, Wickramasinghe equilib. distr. $r_H = 0.012\mu$ | 3597 4348 6100 | .39 .35 .26 | .23 .18 .12 |
| Graphite + dirty ice, core radius $0.03\mu$ equilib. distr. of mantles, $P/\mu = 15.0$ | 3597 4348 6100 | .53 .47 .34 | .51 .40 .21 |
| Platt particles | 3597 4348 6100 | .98 .98 .98 | 0.00 0.00 0.00 |

posite particles, followed by integrations over suitable size distributions. For Platt particles an isotropic phase function and unit albedo were assumed in accordance with Platt (1956).

As was done in the observed intensities, a zero level was introduced in the theoretical intensity distributions at $|b^{II}| = 27°$, thus eliminating the difficulty arising from the fact that the observed intensities have to be considered lower limits. Included in the comparison were particles consisting of dirty ice, graphite, ice-coated graphite, and Platt particles. Grains composed of iron and pure dielectrics could be ruled out beforehand. The parameters $\gamma$ and g used in these calculations are shown in Table 22.

The result of this study was that the particles responsible for the scattered light have an *albedo close to unity* and an *isotropic phase function* in the observed wave length range, properties which are most likely manifested in Platt particles. Graphite and dirty ice can be safely ruled out; ice-coated graphite is unlikely on the basis of Henyey and Greenstein's model.

Confirmation of our work is shown in Figure 13-8 by recent results of Roach at Boulder whose data are represented by the x points and the open

**Table 23. Final Intensities of the DGR.**

| Region | n(U) | | n(B) | | n(ORANGE) | |
|---|---|---|---|---|---|---|
| $b^{II} =$ | 0° | 25° | 0° | 25° | 0° | 25° |
| Cygnus | 61±6 | 20 | 58±6 | 18 | 120±15 | 30 |
| Taurus-Auriga | 42±5 | 14 | 39±6 | 12 | .. | .. |

circles. Our work in Cygnus is shown as filled circles and our Tau-Aur work is designated by triangles. For comparison calculations have been made for the following models:

I   Graphite in the form of plates.
II   Graphite in the form of spheres.
III   Graphite plates with an ice covering.
IV   Graphite spheres with an ice covering.
V   Dirty ice.
VI   Platt particles

The experimental points fit best with the last model.

Table 23 lists the final intensities of the DGR, corrected for the finite contribution of the intensity level at $b^{II} = 27°$. The errors quoted include not only the observational uncertainties but also the fact that different scans covered slightly different areas in the sky.

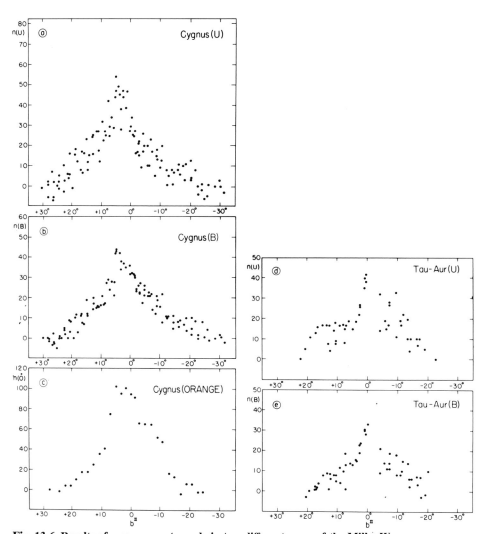

**Fig. 13-6. Results of measurements made in two different areas of the Milky Way.**

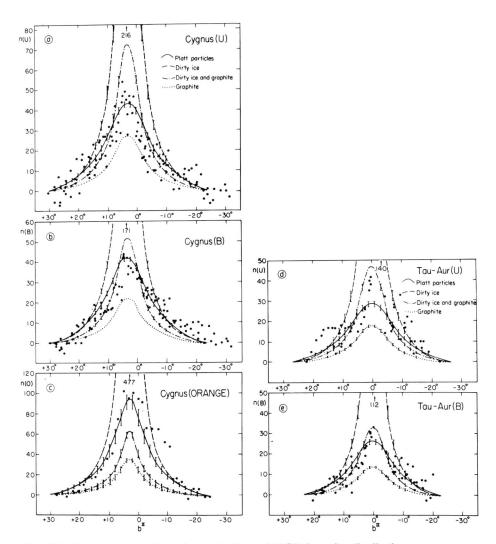

**Fig. 13-7. Comparison of theoretical and observed DGR intensity distributions.**

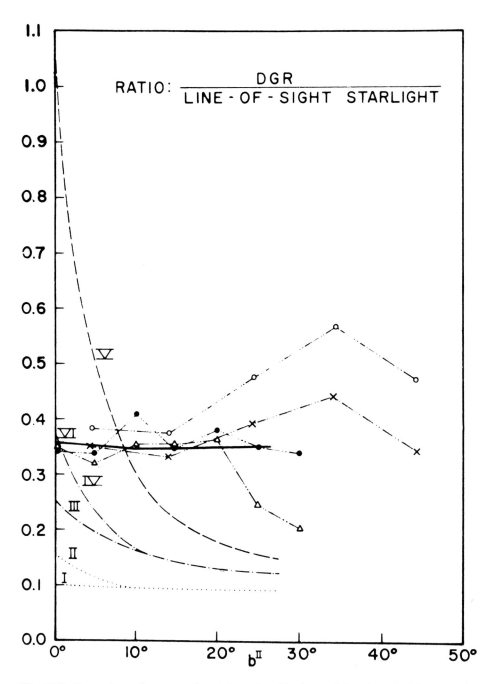

**Fig. 13-8. Comparison of our experimental results with those of Roach and with calculations based upon six models. (See text for details.)**

Our conclusions are strongly supported by the results of a detailed analysis of O'Dell's (1965) photoelectric measurements of the Merope reflection nebula. The radar scattering curve and the observed color excesses can be understood, if the scattering dust grains are Platt particles and the wave length dependence of the scattering efficiency equals the wave length dependence of the interstellar extinction as found in the reddening curve.

Additional work to test the validity of these statements for the general interstellar dust at wave lengths as short as $\lambda 2200$ Å based on rocket measurements is in progress. It is hoped, also, that through systematic investigation of diffuse interstellar absorption features additional information regarding the physical structure and chemical nature of the interstellar grains can be obtained.

SUMMARY

By use of the Henyey-Greenstein model for the radiative transfer of diffuse galactic light, the intensity distribution of this radiation has been predicted for various models of interstellar particles. These predictions have been compared with modern observations in three band-passes. The comparison seems to require an isotropic phase function and an albedo near unity as properties of interstellar dust particles. This seems to rule out models based on Mie theory and supports the Platt model for interstellar dust.

DISCUSSION

(Q) Particles of 20 Å suggest organic molecules.
(A) Witt: Platt was silent about this, but it is assumed that they are made up of C and H; thus they may be material out of which carbonaceous chondrites are made. Organic molecules are found there.
(Q) Has anything been done about ultraviolet scattering by interstellar dust?
(A) Witt: Work is in progress.

# Cometary Nuclei
## ARMAND H. DELSEMME

*Abstract: Cometary nuclei could be the last remaining samples of the early composition of the solar system. The "dirty snowball" model still appears to be the best approach to their understanding. However, the observed lifetimes of the unknown "parents" of the identified radicals cannot be reconciled so far with likely molecules. It is suggested here that these lifetimes could be the lifetimes of the ice grains removed from the nucleus by evaporating gases.*

INTRODUCTION

The chemical composition of meteorites is certainly one of our best sources of *direct* information on extra-terrestrial matter. However, this composition, as observed when they have reached the ground, may have been deeply modified in two respects.

First, they have, of course, been modified by their entry into the atmosphere and this violent heat treatment could have vaporized a great deal of the most volatile materials.

Second, they may have remained for, perhaps, five billion years within the inner solar system before reaching the earth. This second heat treatment, if less obvious, is likely to be a lot more severe than a mere entry through our atmosphere. On small bodies without measurable gravity, no volatile material could remain for eons when they are in a region where the sun's radiation heats them up.

However, there are *other* heavenly bodies, still present in the solar system, which may have kept all their volatile materials as a sample of its early composition, and they give a clue about these volatile materials through their spectra which are therefore very exciting to observe and very important to understand. These are the comets, and my topic today will be the cometary nucleus as a sample of extra-terrestrial matter.

Why are comets believed to be samples of the early composition of the solar system? I will sketch only an incomplete answer in a few words: Oort has described a mechanism which has been highly effective to remove comets from the inner solar system, as soon as they were created as independent bodies, and to put them in a "deep freeze" for five billion years; and a deep freeze is a very effective way to keep snowballs from evaporating. The mechanism lies in the *perturbations* by the planets, mainly Jupiter, and this "deep freeze" is a shell-shaped zone spherically distributed extending roughly from 100,000 to 200,000 A.U. around the sun at mid-distance to the nearby stars, a zone where perhaps one hundred billion comets move at random in orbits still dynamically linked with the solar system and centered on the sun. Occasionally nearby stars perturb one of these orbits and send a comet back on a plunging trajectory toward the sun, where it may be captured again by planetary perturbations and therefore "die" quickly, that is, in a few centuries or milleniums. This short time is enough to evaporate all its gases and to distribute all its nonvolatile materials along its trajectory, making sometimes for our benefit wonderful meteor showers in the earth's atmosphere. The statistical distribution of the semi-major axis of the orbits computed before planetary perturbations or rather their reciprocal, which is proportional to the total gravitational energy, strongly supports Oort's hypothesis.

Let's rather come back to what we really know for sure about comets. When we see a comet far away from the sun, say 10 to 15 A.U., it generally has a starlike appearance; we see only what we call the *nucleus*. As the comet is nearing the sun we see a *coma*, emerging from the nucleus, between 6 and 4 A.U. from the sun. (In the accepted terminology, coma and nucleus together form the "head" of the comet.) The coma is a kind of diffuse, luminous, and nebulous atmosphere with no precise limit (it merges with the sky background). The coma diameter may reach $10^5$ to $10^6$ km. On the contrary, the nucleus remains starlike even when very near the earth; photometry suggests diameters between 0.1 and 10 km. At last, near 3 A.U., one or several tails appear, which may reach $10^7$ to $10^8$ km, about always roughly directed away from the sun. What is the observational evidence given by cometary spectra? At large distances from the sun, when the nucleus is still alone, it gives a pure reflection, most likely given by small particles of dust. But either alone, or superimposed on such a reflection spectrum, almost always appears a fluorescence spectrum, fluorescence excited by the ultraviolet spectrum of the sun and showing emission bands from the following radicals:

| *Heads* | *Tails* |
|---------|---------|
| OH, [OI] | $OH^+$ |
| CH | $CH^+$ |
| $NH$, $NH_2$ | $N_2$ |
| CN | $CO^+$ |
| $C_2$ | $CO_2{}^+$ |
| $C_3$ | |

This atmosphere of molecular fragments is steadily scattered and lost in space. It seems to come from the dissociation of more stable parent molecules that we cannot see, steadily evaporating from the nucleus as it heats up when it is nearer and nearer the sun. The mechanism of dissociation of these stable molecules in the solar field is very complex; it results from interactions of these invisible molecules with solar ultraviolet and solar particles, and it is not yet thoroughly understood. However, whatever the mechanism, it is obvious that the source and the origin of all the phenomena lie in the cometary nucleus. What is exactly known about the nucleus of a comet?

NUCLEI

The first information on cometary nuclei dates back to the nineteenth century, when it was realized that several recurrent meteor showers were associated with cometary orbits. It was indeed positively established in 1866. On November 13, 1866, tens of thousands of meteors were seen coming from the point of the sky where the earth's orbit was just crossing for a few hours, the orbit of *Comet Temple*. When the trajectories of the *Leonid* meteorites were computed, they were strikingly similar to the orbit of Comet Temple, not only as coming at the right time from the right place, but also with the right velocity vector showing that they had been describing the same elongated elliptical trajectory as Comet Temple. Several coincidences of that kind have been observed since.

Table 24 gives the classical comparison of the parameters describing the orbit of Comet Temple and the orbit of the Leonids, as well as a most recent comparison of the orbit of Comet Giaccobini-Zinner with the Giaccobinids (1933-1946). In Table 24, P means the period, a the semi-major axis of the orbit, p its perihelion distance, e its eccentricity, ω the argument of perihelion, Ω the longitude of the ascending mode and i the inclination of the orbit.

In 1866, the trajectory of the Leonids was computed from visual observations only, which gave a computed orbit surprisingly similar to Comet Temple's near the earth. Its extrapolation at large distances from the earth gave, however, a semi-major axis a bit too large and a computed period obviously inaccurate, as shown by the observed periodicity of the Leonids which trustfully had come back in 1799, 1833, and 1866 with exactly the same period as Comet Temple. In 1946, the figures speak for themselves.

**Table 24. Comparison of the Orbital Elements of Two Comets and Two Meteor Showers. (See Text for Explanation.)**

|  | $P$ | $a$ | $p$ | $e$ | $\omega$ | $\Omega$ | $i$ |
|---|---|---|---|---|---|---|---|
| Leonids |  |  |  |  |  |  |  |
| 1799 — 1833 — 1866 | 45y | 13 | 0.970 | 0.91 | 174° | 235° | 162.5° |
| Comet Temple | 33y | 10 | 0.970 | 0.92 | 171° | 231° | 162.7° |
| Giaccobinids |  |  |  |  |  |  |  |
| 1933 — 1946 | 6.59y | 3.51 | 0.996 | 0.717 | 171.8° | 196.2° | 30.7° |
| Comet Giaccobini-Zinner | 6.59y | 3.51 | 0.996 | 0.717 | 171.8° | 196.2° | 30.7° |

Not *all* the known meteorites come from comets; we have heard yesterday that many of them are probably fragments of larger bodies, but *all* the *shower* meteors seem to come from comets. It is therefore not very surprising that, for many years, there was a general consensus on the fact that a cometary nucleus was nothing other than a cloud of meteorites. Hence, the origin of the so-called "sand bank" model of the cometary nucleus, which describes the nucleus as a loose agglomeration of meteoritic particles, carrying with them a minor amount of gas, for instance adsorbed in the pores of the meteorites and providing for the activity of the head and the tail under a solar activation. However, the fact that comets do exist implied a minimum cohesion for the nucleus. The agent of this cohesion was first thought to be gravitation. Disintegration of comets were explained as arising from solar heating or tide-raising effects.

It soon became apparent, however, that the argument of the gravitational cohesion had to be abandoned. Because of the very small masses involved in a cometary nucleus, gravitational cohesion could work only for densities several orders of magnitude higher than meteoritic densities with neither any possible mechanism to reach these densities now nor any argument to explain them in the history of comets.

If gravitation is not involved, what is the cement linking together the meteoritic material in the cometary nucleus? It is most likely a volatile substance.

ICES OF FROZEN GASES

From these ideas and from different considerations, in particular from orbit perturbations coming from losses of mass, Whipple suggested in the 1950's that the *major* constituent of a cometary nucleus was in fact ices, ices of frozen gases, and that these ices could constitute an important, if not, predominant fraction of the mass of, at least, "new" or undissociated comets.

When I say ices, I would rather say *snows*, not only water snow, but also methane snow, ammonia snow, etc.—snows of frozen gases of many volatile substances with stony meteorites imbedded in them. The cometary nucleus would be, as Whipple puts it, a "dirty snowball" of a few miles' diameter. This "icy conglomerate" model escapes the major difficulties of the sand bank model, and explains many odd peculiarities of comets. The mere existence of these snowballs for five billion years is no more an enigma if they *do* come from the deep freeze suggested by Oort's cloud.

However, the Whipple model of the cometary nucleus did not quantitatively solve the whole problem. On the contrary, it was faced, in its original shape, with a major difficulty. Sublimation rates in vacuum of $H_2O$, $CH_4$ and $NH_3$ pure ices have extraordinarily different orders of magnitude. For instance, around $100°$ K., for each gram of water vapor produced there are 10 tons of $NH_3$ vapor and 100 million tons of $CH_4$ vapor produced. One sees at first glance that these rates don't fit in with the intensities observed in the cometary spectra for OH, NH, and CH.

To put it another way, production rates would be the same, for $H_2O$ at 1.5 A.U. from the sun, for $NH_3$ at 3.8 A.U., and for $CH_4$ at 19 A.U.; this does not fit at all with the distances where the cometary bands appear. Aware of these difficulties, Swings and I published a small note in 1952, suggesting to explain the situation by the fact that the cometary ices were not pure ices of each substance considered, but that they were *water hydrates* of $CH_4$, $CO_2$, $NH_3$, etc. Of course the water hydrate of $NH_3$ is the ionic hydrate $NH_4^+$ $OH^-$; solutions of ammonia in water are well known. The other hydrates are clathrate hydrates. If gases are tied into the hydrate structure, the rate of vaporization of the water linked within the hydrates is not notably changed, but the rates of $NH_3$ and $CH_4$ are woefully reduced.

The idea was acceptable, as it was confirmed by S. Miller in 1961. Miller's contribution was outstanding because he carefully studied dissociation pressures of hydrates in the laboratory near $200°$ K., which gave a strong experimental basis for extrapolation toward lower temperatures.

Of course, the hypothesis of the occurrence of the gas hydrates is likely, only if there is no doubt about the presence of the water molecule within the cometary nucleus. Whipple (1950) had already suggested that some radicals could permanently exist in cometary nuclei. Haser (1955), developing this idea, proposed to consider that the observed OH comes from OH radicals permanently stored inside the nucleus, $H_2O$ becoming useless in this case as a parent molecule of OH.

PRESENCE OF WATER

The idea was that OH is seen too far away from the sun to come from $H_2O$, water having too low a vapor pressure. This idea had never been checked quantitatively, but if it were true, it could ruin the whole model imagined by Whipple. Besides, it could become a crucial argument to distinguish between two possible origins of comets.

Three years ago, I was able to show that this argument was not true. I compared the mean value of OH production observed in a comet, taken from data published in the literature, with the water vapor production computed from the evaporation theory, and it is obvious that there is *certainly enough* water to explain OH by the dissociation of $H_2O$. In fact, from the rates of dissociation of water vapor published by Bierman, it appeared that I had overshot my target by a factor of 200 to 500, which was a bit too much to be comfortable. There was no longer not enough water to explain OH, there was too much of it! The matter was settled very recently when I became aware of Elisabeth Roemer's new statistics not yet published on the geometrical dimensions of the cometary nuclei. At last we have good meaningful values for the radii; my model had a radius at least ten times too large, which means that the evaporation rate, which is proportional to the surface of the nucleus, was two orders of magnitude too high.

Now the orders of magnitude of $H_2O$ production from evaporation theory and of OH concentration from observational data are no longer wide apart, and there is no reason any more to doubt that the most probable source of OH is $H_2O$. As the argument for the presence of water now stands on an almost quantitative basis, this implies the presence of hydrates, because of the thermodynamics of the reaction. I feel, therefore, safer now in discussing the general theory of gas production and the model of cometary nucleus that I have recently pubished and which brought for the first time this almost quantitative agreement with the observations.

I accept first a simplified spherical Whipple model of the nucleus, with a homogeneous crust that I visualize as made of an open structure of small crystals in needles linked by filamentary structures rather than blocks of ices; second, I accept the existence of a steady state which implies a balance, at each point of the cometary crust, between the heat flow coming from the sun and the heat flow leaving the comet. This means in particular that the rotation of the nucleus about its axis is slow enough. This steady state is of course slowly shifting with the variation of the solar distance of the comet.

The heat flow entering the comet's nucleus is the part of the solar flux which is not reflected. The heat flow leaving the comet comes from the infrared radiation leaving the nucleus, given by Stephan's law, plus the latent heat of the sublimation of the snows, or rather in a more general way, the free energy changes coming from any chemical reactions of physico-chemical phenomena (mere evaporation as well as adsorption) happening within the surface of the nucleus and liberating gases.

The major interest of the equation is to be computable numerically for almost any model, because there are tables giving entropy and enthalpy for a wide range of temperatures between $50°$ K. and $1000°$ K. for many physico-chemical changes and many chemical reactions, for most of the usual molecules.

MODELS

The next problem is choosing models for computing numerical values of this equation in some case studies. To start with, I have accepted cosmical abundances of carbon, nitrogen and oxygen, and the presence in the early times of a great excess of hydrogen whenever the comet began to build up its nucleus, excess which was lost later with helium and the most volatile gases after having hydrogenated everything possible. Cooling down such a cosmical mixture, the three most abundant molecules by far are of course $H_2O$, $CH_4$ and $NH_3$. $CH_4$ and $NH_3$ are then hydrated by water and it comes:

$$
\begin{array}{l}
12\% \ CH_4 \\
62\% \ CH_4 \text{ hydrate} \\
12\% \ NH_3 \text{ hydrate} \\
14\% \ Fe,Mg,Si,S \ldots
\end{array}
$$

the last 14 per cent being the usual meteorite fragments. Within a plausible range of temperatures the whole $NH_3$ and part of $CH_4$ are necessarily stabilized into hydrates: it is unavoidable because of chemical thermodynamics. The methane excess is likely to be lost very quickly, and I showed that it is usually so in less than one or two perihelion passages. The peculiar behavior of "new" comets could be explained in terms of vaporization of free methane, if there is any, at specially low temperatures. If one compares this model with the observed photometric curves of Comet Arend-Roland, one comes to the conclusion that Comet Arend-Roland could be a fair example of a comet with a methane excess; vaporization heats: 6.800 cal/mole for $H_2O$ or hydrates, 2.600 cal/mole for $CH_4$, about 2.000 $\pm500$ cal/mole for Comet Arend-Roland mean vaporization heat.

As a second case study, a model without any methane excess was computed, as representative either of different genuine abundances or as describing a comet having already made at least one perihelion passage. In this case no stratification is likely to build up from gravity because crystal densities are very similar for present hydrates. The steady state equilibrium is reached very rapidly and evaporation rates build up in such a way that, within $\pm1$ per cent, Stephan's Law *alone* gives the surface temperature for distances greater than 6 A.U., and vaporization rates *alone* give the surface temperature for distances smaller than 0.5 A.U.; the intermediate distances between 0.5 and 6 A.U., which are precisely the most interesting ones, define the range

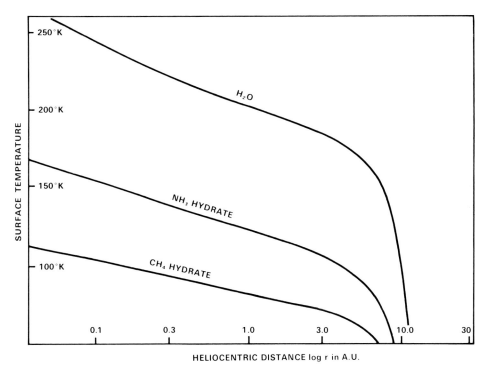

Fig. 14-1. Surface temperatures of comets as a function of distance from the sun for three different models.

where the two terms are exchanging their orders of magnitude. These surface temperatures are given in Figure 14-1. They are still rather cool when the comet is approaching its perihelion: 200° K. at 1 A.U. from the sun, 250° K. at 0.1 A.U. from the sun. In particular, fusion temperatures will never be reached even for ammonia eutectics which have low fusion temperatures. This means that snows will remain dry after perihelion. This is an important remark as far as albedo is concerned.

COMA

Sublimating gases spreading out into the coma keep about the same composition throughout the comet's course, roughly fixed, as far as *methane* is concerned, by the existence of the *hydrates,* and as far as *ammonia* is concerned, by the cosmical abundance of *nitrogen.*

$$\boxed{H_2O:\ 80\%\,;\ CH_4:\ 12\%\,;\ NH_3:\ 8\%}$$

As one knows surface temperatures, the mean velocity of the kinetic theory is defined and one can compute, from molecule production, the number $N_i$ per volume unit at the level of the nuclear surface as well as the mean free path. If one first neglects any outside interference as from solar ultraviolet or particles, one can compute a coma radius, versus solar distance, by arbitrarily choosing a definition providing the same optical depth at the boundary of the coma whatever the radius.

Figure 14-2 shows that this total coma does not exist at 6 A.U., appears suddenly around 5 A.U., is 1000 km deep at 3 A.U., 10,000 km at 1 A.U. and 200,000 km at 0.3 A.U. This model is therefore surprisingly successful in predicting the right range of apparition of the coma.

We have also defined a radius limiting the central part of the coma: where molecular collisions are still frequent and therefore where chemical reactions in gases are likely to proceed very far toward their equilibrium. Figure 14-2 shows that this chemical coma appears *suddenly* around 4 A.U. At 3.7 A.U. its thickness is only 10 km above the nucleus, 70 km near 1 A.U. and 300 km near 0.3 A.U.

As the coma is mainly formed with water vapor, its highest layers strongly absorb solar light, mainly in ultraviolet, and thus heat up quickly. The way the heat wave enters the coma is best described (Figure 14-2) by the altitude of the 1500° K. isotherm. The isotherm which has not yet entered the coma for

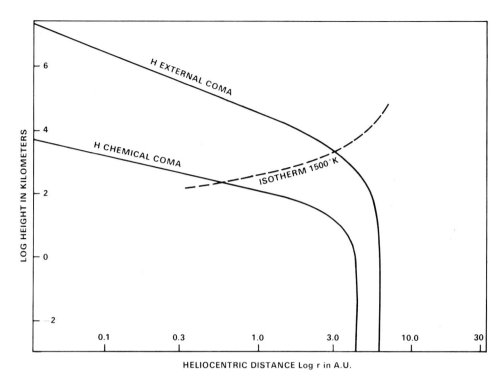

**Fig. 14-2.** Calculated height of coma as a function of comet distance from the sun.

**Table 25. Pyrolysis Caused by Heating Gives Progressive Dehydration of the Coma.**

| Reactants | Equilibrium | Observed | |
|---|---|---|---|
| | (1500° K) | Heads | Tails |
| 80% $H_2O$ | 68% $H_2O$ | OH, [OI] | $OH^+$ |
| 12% $CH_4$ | 14% $H_2$ | — | — |
| 8% $NH_3$ | 8% $CH_4$ | CH | $CH^+$ |
| | 5% $NH_3$ | NH, $NH_2$ | — |
| | 2% CO | — | $CO^+$ |
| | 1% $CO_2$ | — | $CO_2^+$ |
| | 1% $N_2$ | — | $N_2^+$ |
| | 1% HCN | CN | — |
| | * 0.1% $C_2H_n$ | $C_2$ | — |
| | * 0.01% $C_3H_n$ | $C_3$ | — |

more than 3 A.U., enters the outside coma near 3 A.U. and enters the chemical coma near 0.5 A.U.

This rough computation is sufficient to convince me that this heating may lead to pyrolysis, giving a progressive dehydrogenation which Table 25 illustrates.

THE HYDROGEN PROBLEM

Notice that all these reactions do liberate free hydrogen. Why doesn't one see it? The hydrogen atom could be seen only through a fluorescence mechanism, involving the excitation by the solar Lyman $\beta$, whose intensity is known now from measurements out of the atmosphere; the Balmer $\alpha$ line excited through this mechanism is just too weak to be detected in comets. It is, however, worth noting that, apart from $H_2$, each molecule obtained might be the parent of an observed radical (see Table 25, last column, at left for the coma, at right for the tail), and has a possible parent molecule, no other molecule being needed whatsoever to explain observations in head and tail. This double correspondence is, of course, of a qualitative nature only. First, there are no good observational data, and second, the thermodynamic equilibrium is certainly still far away from the coma conditions.

Therefore, the next step would be to fit a model of the cometary nucleus with a model of the coma. The goal of such a study would be the quantitative explanation of the origin of the radicals observed in cometary comas. The simplest idea which was repeatedly tried is to suppose that these radicals arise from the dissociation of parent molecules sublimating from the nucleus without any intermediate step such as pyrolysis. In this hypothesis, the lifetimes of the parent molecules could be deduced from a few photometric profiles of the coma measured in monochromatic light, as by Malaise (1966), if the dissociation mechanisms were understood. However, the known photometric profiles do not extend far enough and only give orders of magnitude or upper limits for the molecular lifetimes. Despite this, there are such large discrepancies in the lifetimes of assumed molecules that it suggests that other mechanisms must be found. In this respect, I think that dissociation should be applied to by-products of a pyrolysis of the parent molecules, this pyrolysis taking place by solar heating of the inner coma. This idea that I suggested repeatedly in 1952 and in 1966 has never been checked in a quantitative manner. It could be developed in a tentative model along the following lines. Some

**Table 26.**

| Suggested Pyrolyses | Subsequent Processes |
|---|---|
| $NH_3 + CH_4 \rightarrow HCN + 3H_2$ | $HCN + h\nu \rightarrow \underline{CN}$ observed |
| $H_2O + CH_4 \rightarrow CO + 3H_2$ | $CO + h\nu \rightarrow \underline{CO^+}$ observed |
| $H_2O + CO \rightarrow CO_2 + H_2$ | $CO_2 + h\nu \rightarrow \underline{CO_2^+}$ observed |
| $2NH_3 \rightarrow N_2 + 3H_2$ | $N_2 + h\nu \rightarrow \underline{N_2^+}$ observed |
| $2CH_4 \rightarrow C_2H_2 + 3H_2$ | $C_2H_2 + h\nu \rightarrow \underline{C_2}$ observed |
| $3CH_4 \rightarrow C_3H_4 + 4H_2$ | $C_3H_4 + h\nu \rightarrow \underline{C_3}$ observed |

of the radicals observed could still come from abundant parent molecules without any intermediate steps. For instance:

$$H_2O + h\nu \rightarrow OH, OH^+, O$$
$$CH_4 + h\nu \rightarrow CH, CH^+$$
$$NH_4 + h\nu \rightarrow NH, NH_2$$

But the other radicals present in comets could stem from intermediate molecules produced by pyrolysis of the parent molecules as in Table 26:

A careful analysis of the physico-chemical equilibria that may take place in the "chemical coma" is therefore worthwhile to attempt. The results of such an analysis could be incorporated into a model of the coma and, hopefully, could remove some of the discrepancies existing in the lifetimes of assumed molecules, as well as explain all the observed features from three parent molecules: $H_2O$, $CH_4$, and $NH_3$ only. One of my students, D. Miller, is now working on this particular problem.

Some observations (Dossin 1962) suggest that the chemical coma could have a non-negligible optical density, and therefore it is also proposed to investigate the radiative transfer effects of such an optical density on the whole model, in particular its meaning for the evaporation theory of the nucleus. Some simple mathematical models can be computed, for example the homogeneous isotropic scattering of light in the chemical coma by a fog of ice or snow particles. Huebner (1966) has recently published an interesting paper along these lines.

Urey and Donn (1955) and Donn (1960), after suggesting the existence of trapped radicals as a major cause of cometary phenomena, seem to have now dismissed this hypothesis. I am not aware, however, that a careful discussion of the arguments involved has been published.

To sum up the situation, we have already an idea of what may happen in the inner core. We are also beginning to get some models of coma, explaining some dissociations (or ionizations for the tail) in terms of solar ultraviolet or particles.

To link nucleus and coma phenomena, the basic idea of the vaporization process was already clear in Whipple's model.

However, there are missing links between the molecules evaporating from the nucleus and the radicals and ions observed in the coma and tail. The parent molecules are still hypothetical and the mechanism of their dissociation and ionization is not at this time quantitatively understood, as expressed by Wurm (1962), Wurm and Mammaus (1966), Potter and Del Duca (1964), and Wurm (1966).

If we are unable to reconcile observed lifetimes with possible lifetimes of likely parent molecules, I want to suggest here that we could substitute lifetimes of ice grains undergoing evaporation within the coma, instead of lifetimes of parent molecules, to explain the peculiar spectrophotometric gradients that are observed in cometary heads.

My personal conviction is that the surface temperature of the cometary nucleus is the clue linking inner and outer phenomena, as well as, for instance, in stellar atmospheres. In this respect, remember that in the 1920's, the first successful theory of ionization in stellar atmospheres was built up in terms of temperatures; temperature explained everything in the spectral sequence of stars; it explained the kind of equilibrium reached to define partial pressures of ions and electrons, and it helped later to fit in models down to the inner core of the stars. In the same way, it is my strong belief that surface temperature will be at the core of our understanding of the cometary nucleus, which could likely be one of the most ancient samples of extra-terrestrial matter still available in the solar system.

## DISCUSSION

(Q) Do I infer from Table 1 that in 1866 people didn't recognize the association with the 1833 and 1799 meteor showers?

(A) Delsemme: No. Visual correspondence was known, but only later were the calculations of the orbital elements made with sufficient accuracy.

(Q) What is your feeling about how much lower is the temperature of the core of the nucleus than the temperature of the surface?

(A) Delsemme: Very much lower indeed. It must still be at the temperature it had when in Oort's cloud.

(Q) What is the basis of your assertion that most of the methane would be lost during the first few perihelions?

(A) Delsemme: Through the separation mechanisms of crystals rather than thermal separation. In small gravitational fields and with small cohesions I have discussed these mechanisms. The "dry ice" is loose and small density differences are important. For example, $\rho = 0.44$ for methane crystals, $\rho = 0.92$ to 0.96 for water.

(Q) I noticed no $CH_2$ or $CH_3$ in your list of observed radicals.

(A) Delsemme: None is observed so far in cometary spectra.

# VI. Conference Summary

# Panel Discussion
## JOHN A. O'KEEFE, CHAIRMAN;
## JAMES H. PATTERSON;
## CHARLES A. RANDALL;
## EDWARD ZELLER

O'KEEFE: The question to which we shall address ourselves is research projects which might be suitable for universities whose means are not at the top level but which are by no means diminishing. We shall treat this problem from several different viewpoints, then we shall be open to questions from the floor. I shall ask Professor Randall to lead off.

RANDALL: We have discussed several problems at a very high scientific level during the course of this conference. At the outset it should be said that the universities represented here are not necessarily small nor are they necessarily young. The common denominator seems to be that their graduate programs are relatively recent. We have been called burgeoning universities. Perhaps one purpose of this conference is to find the hormone which will bring us to full flower.

Two problems can be identified from corridor conversations as well as from the excellent technical papers. First, all too often there is insufficient communication between the burgeoning universities and the outstanding scientists of our country. I can see that a given person with a good deal of ability and enthusiasm and with a minimum of capital equipment might have excellent ideas, but would have some misgivings about embarking upon a particular research project because some of the larger university groups and federal laboratories are already working on the problem and might beat him to it. Conferences of this sort will help to solve this communication problem. One may say, "Why doesn't he merely keep up with the literature?" but information retrieval is itself becoming a major problem today. Literature is always history. What a worker really wants to know is what is someone like Professor Meyer planning to do in the next year. We had an excellent example this morning, and he discussed his new experiments freely after the meeting, thus helping to solve the problem of communication.

Second, one of the aims of the CSUI is to encourage inter-university cooperation. A committee of the CSUI has written a proposal for an interesting research problem in solid state. It has no federal support yet, but it has one basic virtue, namely, each of the four universities whose representatives got together to draft the proposal has certain specialized equipment of good quality. But in no case could any one of the universities tackle the over-all problem as formulated by the group. Since the equipment was not duplicated they were able to come up with a proposal in which the problem was identified and each of the four groups could solve a part of the problem. This is not the same as fragmentation into four little separate problems in order to make a big one which would appeal to the federal dollar.

It would be my hope that out of a conference of this sort we might explore avenues of further inter-university cooperation and better communication.

ZELLER: I have the list of major items of equipment submitted by each of your representatives, after the first session. I see NMR, ESR, mass spectrometers, accelerators, electron microscopes, x-ray diffraction, etc. We are not dealing with small-time operations. Any sharing among institutions would help greatly and would impress federal agencies.

My principal plea is for cooperation within a university. The federal agencies and the university administrations all talk about it, agree that they want to support it, but nobody does anything about it. Why don't they? Because joint appointments between departments do not seem to work well in universities, and joint proposals are difficult to steer into the right support channels in federal agencies. I am sorry that this is true. Lip service is paid but no dollars. We must keep pushing. People who had labels of "physicist" may end up working in geology, and conversely. There is no reason why this should not be true. Your universities probably have more flexibility in this effort than is found in some of the more prestigious universities.

You seem to have equipment to do almost anything you want to do collectively. There appears to be adequate equipment.

O'KEEFE: I am going to take issue slightly with my colleagues and offer some research projects which are not too expensive although they are not of the sealing wax and string type. I heard some discussion about the prospect of looking at erosion on the lunar surface or on planetary bodies. Shoemaker is a great man; he is a friend of mine, but many statements he made about erosion would not be unequivocally acceptable and it is my opinion that some of his statements were wrong. That is not to stay that he hasn't done the best anyone could do under the circumstances, but I am not sure that what he called a regolith is in fact a debris layer. It looks more like volcanic ash. I am not quite sure that the mechanism of erosion he discussed is consistent with what we know about the rate at which meteorites are eroded. I think that it might be very much worthwhile to re-do this whole problem of the erosion of the lunar surface. If you want to do it on a string and sealing wax basis, all you have to do is ask the space agency for the Lunar Orbiter and the Surveyor photographs. They are available from the Data Center at Goddard. There are about 5,000 photographs with fantastic detail—far more on one of these pictures than on all the previous ones put together. These have not been fully exploited. Only rather superficial examination has been made. People from a small university can get these and go after the problems of geology and erosion from a more analytical standpoint than the geologists have done in the past. This offers promising prospects to the young and hopeful to get their bruises in an honorable way. It will not be easy, but the inter-disciplinary field is where things really get done. If a physicist or a chemist goes into geology to talk about geology the geologist will not be pleased; however, it is essential that he do so because geologists are no longer able to handle the problems of planetology as they did the problems of geology. They dealt almost exclusively with words, hardly ever with numbers, and seldom with physical laws. This is not a reasonable way to approach the problem of the surface of the moon. The parameters can be varied.

It may be possible to do better than that because if the analyses reported by Dr. Patterson are indeed correct then it is very likely that they are related to the basaltic achondrites. If we combine them with the lunar pictures it is possible to draw some very important conclusions. For example: In the pictures we often see rocks which are more or less rounded, hemispherical. In a few cases where these rocks were flipped over by the gas jet or the digger we find on the bottom side the rock is quite rough. This same thing is found in the eucrites, the basaltic achondrites or lithic fragments—hemispheric on one side and rough on the other. Such a property is the result of mechanical erosion.

One can get started in meteorites by purchasing from any of a number of good dealers. I have the names of many in my files.

An idea which has fascinated me is related to a meteor reported about fifty-five years ago. It started in Alberta and ended in the South Pacific. It was moving in a path which was approximately parallel to the surface of the earth. It was thought then and is now believed to have been a natural satellite of the earth. You will find this mentioned in no astronomy textbook, but the publications are by reasonable, responsible people; the basic publication is by Donald A. MacRae, who later became the director of the David Dunlap Observatory. It is a beautiful piece of work. Why do we let a thing like this pass by? Why not go into upstate New York where the thing passed overhead and look for micrometeorites and tektites in the lakes? We know exactly where to look.

Another basic problem is the lunar orbit. For hundreds of years astronomers and others have done an incredible amount of work on this problem. Recently the geologists, biologists and malacologists [malacology deals with mollusks] have started to work on it. On the outside of clam shells there are striations which result from diurnal deposits of calcium. The thickness of the striations depends upon a number of parameters such as tide height, temperature, etc., so that there is a diurnal cycle, a monthly cycle, and an annual cycle in the striations. Thus, it is possible to ask the clam to tell us the number of days per year; the result in recent times is 365 which is not astonishing. For paleontological specimens we find more than 365. Such studies have been carried back 4 to $5 \times 10^8$ years in clams and corals. It is a new subject and easy to get into.

What is an interesting subject—a subject which we care about in space science? The answer is: any subject which will tell us about cosmology—about the beginnings and the ends of things. (In space science we obtain more information per fact discovered than in other sciences.) I am quite aware that there are theological implications in such studies but I shall not discuss them.

We should also take a very careful look at the volatile materials in tektites. This requires more careful work but is not too expensive. We should study the chemical compositions of tektites because it could tell us how they were formed and in turn how the moon was formed, if they have any relation to the moon. It is conceivable that, from studies of element abundances, the depletion of soda and potash in the basaltic achondrites and the tektites may indicate that the moon was formed at a time when it, too, was depleted in soda and potash. For example, suppose it was formed by fission from the earth. This would have been a catastrophic process which would certainly affect the volatiles in tektites.

Another important study is impact craters on the earth. Somebody should study very accurate maps (such as those made by oil companies if you can get hold of them) and count the number produced per square kilometer per year.

Geodesy is another area of interest. If we go way back into history we realize that the space age is now about ten years old and during that time we have made incredible progress in finding how non-spherical the earth is. For example, the flattening of the earth is known to five significant figures. It is only about 20 km which means we know the earth's diameter to within 6 in. Who really believes that the surface of the earth is constant within 6 in. over a period of ten years? If the leaves fall off the trees things change that much. Therefore one important problem is time variation of geodetic parameters.

We have heard discussions of a solar nebula postulated to exist long ago from which the planets were formed by making "big pieces out of little ones." Physically, the converse is much more likely; it is easier to make little ones out of big ones. If that is true the study of binary stars assumes importance in the evolution of our own solar system.

Many, many more observations of the details of the moon's surface should be made from the earth. For example, systematic work of this sort could give valuable information about the luminescence problem.

PATTERSON: From my limited experience with NASA and other federal agencies I offer a bit of advice to someone who has an idea that might be used in the space program or any other large program. It is: get into the business early. Seven years ago I made a proposal to NASA and was told we were already too late; the program was fixed. The second bit of advice is: don't be discouraged by such talk. It was fixed at that time, but things change. We did get into it. The third bit of advice: be ready to change your ideas and modify your systems to go along with changing conditions. The Surveyor program changed, the rockets proved to be different, other experiments fell off, and we were included.

CASELLA (Northern Illinois University): Referring to O'Keefe's suggestion about lunar erosion studies, I tried to obtain the photographs from NASA and was told that it would cost $3,000 for Lunar Orbiter I and II.

ZELLER: We have a three-hour seminar for graduate geologists; normally after two hours they want to leave. Then we started to use the Orbiter photos, and now we never stay less than four hours; we can't drag the students away from the photos. They have unbelievable detail. The students and faculty argue about their interpretation until all hours of the night. We were fortunate to get about 150 without cost.

KANE (Ball State University): Would the geological map of Kentucky be a promising source in which to search for astroblems?

O'KEEFE: A good many have been found this way, but I was referring to sub-surface features found by oil company drillings in the Lower Mississippi Valley —the Gulf States area. They have drilled over an astonishingly close network.

ZELLER: Almost everybody here has access to x-ray equipment and could look for shock-metamorphosed minerals.

O'KEEFE: One very useful problem is to look for the planar features caused by shock in quartz. They have an appearance similar to cleavage planes, although quartz normally has no cleavage planes. A book on this subject will be out shortly.

BURGESS (Ball State University): What height, size, resolution, and magnification are necessary to make solar luminescent studies?

O'KEEFE: It has been done with 12 in. to 15 in. telescopes or smaller. However, a special device developed by Edmund of NASA helps a great deal. It is a moon-blink apparatus which puts the light into two colors. The eye is more sensitive to flicker than to color, so the two images are viewed alternately on a non-color TV screen.

KANE (Ball State University): At some of our universities we have no viable schools of research so we are encouraged to go away for three months to "do research" and then come back and teach "all the new facts." But at many of the large research centers three months is insufficient time to complete a re-search project. Can we come to Goddard on a short-term basis?

O'KEEFE: We have a fellowship program which provides full-time study for the summer.

KASTNER: Argonne National Laboratory has several programs for continuing education. You make direct contact with one of the divisions or you may oper-ate through the Office of College and University Cooperation. I am sure similar programs exist at Oak Ridge, Brookhaven, and other places.

HAMILTON (Ohio State University): There will be a symposium on Continental Drift next week in Columbus.

WALLACE (University of California, Berkeley): The nuclear emulsion technique is an excellent tool for a research program with limited resources. For example, we looked at $10^6$ iron tracks in one month with the aid of only three scanners.

O'KEEFE: Price, Fleischer and Walker have had enormous success with their etched tracks. All you need is HCl and a microscope. They dated tektites, looked for extinct 244, obtained ages of many samples. Electron microscopes can also be used to examine such tracks.

KASTNER: I just bought some emulsions loaded with Li and B for $20.00 per dozen.

KANE (to Kastner): I assume that the neutron bombardment you have measured would change the surface chemistry and isotopic abundance on planets and the moon.

KASTNER: Such changes surely take place.

PATTERSON: Our preliminary data have rather large errors, but it appears that the neutron flux has caused little change because the surface soil appears to be the same chemically as that a short distance below the surface.

PETERSON: I should like to mention the CSUI PACE program financed by the NSF. This provides a semester or a quarter for a faculty member to come to Argonne National Laboratory. Argonne also has similar direct opportunities. I would be glad to discuss details with anyone interested.

# Authors

CHARLES A. RANDALL, of the Department of Physics, Ohio University, Athens, Ohio. Dr. Randall, born in Florida, received the A.B. from Kalamazoo College, the M.A. from Cornell University, and the Ph.D. in physics from the University of Michigan. Before joining the faculty at Ohio University, he taught at Allen Academy in Texas and Wayland Junior College in Wisconsin. He worked as an industrial research physicist for Fairbanks Morse & Co. and was a radar officer in the U. S. Navy. He has also held appointments as a National Science Foundation OEEC Fellow at CERN Laboratories, as research participant at Oak Ridge National Laboratory, as instructor at Goodyear Atomic Corp., member of the Nuclear Emulsion Institute at the University of Chicago, President of the Ohio Academy of Science, and President of the Board of Directors of Central States Universities, Inc. His published papers are mostly in the field of cosmic rays. He is a member of four scientific societies and is a Fellow of the American Physical Society. After twelve years as Chairman of the Department of Physics at Ohio University, he is currently on sabbatical leave doing research in cosmic rays at Sandia Laboratories.

BRIAN MASON ("Composition of Stony Meteorites"), Curator and Supervisor, Division of Meteorites, Smithsonian Institution, United States National Museum, Washington, D. C. Dr. Mason, who has come from New Zealand by way of the Smithsonian Institution, is, because of his enormously important contributions in the study of meteorites, one of the leading scientists in the field. He is a geologist and a mineralogist. His publications are innumerable; besides regular publications and journals, he has written books on geochemistry and meteorites. Perhaps one of the most revealing and important personal characteristics of Dr. Mason is that although he is a museum curator, at the same time he has been a very active research scientist. His role of leadership at the Museum has not been merely the passive one of custodian and collector of these objects that have come to us free from space, but also an active one to investigate these objects and find out more about the region of space in which we exist.

MARTIN H. STUDIER ("Origin of Organic Matter in Meteorites"), of the Chemistry Division, Argonne National Laboratory, Argonne, Illinois. Dr. Studier, born in South Dakota, has been a member of the staff at Argonne for a number of years and has been involved in various programs there even before he received his Ph.D. in chemistry from the University of Chicago. His production rate is surprising in that he has spent about ten years doing things which have not yet been declassified and therefore are not publishable. A recognized authority in the chemistry of heavy elements, Dr. Studier has recently

*326*

rejoined the field of publishable research, and he has become interested in the time-of-flight mass spectrometer. He has made very significant modifications to this instrument and, as a result, was the person best qualified to begin here the detailed investigations of organic matter in meteorites.

JOHN A. O'KEEFE ("Origin of Tektites"), of the Laboratory for Theoretical Studies, National Aeronautics and Space Administration, Goddard Space Flight Center, Greenbelt, Maryland. Dr. O'Keefe, an astronomer who earned a Chicago Ph.D. after graduating from Harvard, is very much interested in matters having to do with the moon, meteorites, and tektites as well as the earth. Born in Lynn, Massachusetts, he has established competence as college professor, mathematician, member of the U. S. Lake Survey, and authority on variable stars. He is one of the most incisive and most stimulating thinkers with respect to many of the current problems in various fields such as meteorites and lunar geology. He is a member or fellow of at least five scientific societies. He has edited a book on tektites, and he has made many contributions to the understanding of these objects which, of all those objects that have arrived on earth, are among the most difficult to explain.

EUGENE SHOEMAKER ("The Lunar Regolith"), of the Astro-Geology Branch, U. S. Geological Survey, Flagstaff, Arizona. Leaving Los Angeles, the city of his birth, he was graduated from California Institute of Technology and earned both the M.S. and Ph.D. from Princeton, where he was a Libby Fellow. He is a specialist in salt anticlines, uranium geology, and the structure of the entire Colorado Plateau. Dr. Shoemaker is the Chief Scientist at Flagstaff. He organized it in 1961 and since that time has done a great deal of research work in what are called astroblems or impact craters on the crust of the earth. He has written some of the most definitive papers on this problem. He is a member of the International Astronomical Union and nine other scientific societies. Of particular interest is the recent announcement that Dr. Shoemaker will be the Chairman of the Department of Geological Sciences at California Institute of Technology, starting in the spring of 1969.

JAMES H. PATTERSON ("Chemical Analysis of the Lunar Surface"), of the Chemistry Division, Argonne National Laboratory, Argonne, Illinois. Born in Kansas City, he was graduated from the University of Omaha and earned the Ph.D. in chemistry at Iowa State University. His fields of specialty include inorganic chemistry, polarography, complex ions in solution, uranium chemistry, and lunar and planetary analysis.

EDWARD ZELLER ("Luminescence and Chemical Effects of Solar Protons Incident Upon the Lunar Surface"), of the Department of Geology, University of Kansas, Lawrence, Kansas. After leaving Peoria, Illinois, he received the A.B. from University of Illinois, the M.A. from Kansas, and the Ph.D. from the University of Wisconsin. He has held fellowships at Berne, Brookhaven, and in the Antarctic. He is a recognized authority in thermoluminescence of geological materials, geological age determination, electron spin resonance and radiation damage in minerals, chemical reactions of high energy protons, and lunar and asteroidal weathering induced by the solar wind.

FRANK B. MCDONALD ("Chemical Composition of Galactic and Solar Cosmic Rays"), of the National Aeronautics and Space Administration, Goddard Space Flight Center, Greenbelt, Maryland. Born in Columbus, Georgia, he received the Bachelor's degree at Duke University, the M.S. and Ph.D. in physics at the University of Minnesota. His first full-time assignment as a professor was at the University of Iowa at Iowa City, where he served for six years in association with Van Allen, whom we all know for his interesting discoveries in space. Dr. McDonald then joined the NASA Goddard Space Flight Center where he became Head of the Fields and Particles Branch in 1959. In the same year and in following years he was associated as project scientist with Explorers 12, 14, and 18. Since 1964 he has also held a joint faculty appointment at the University of Maryland. He is directing some students in advanced degrees—a very wholesome relationship for a man in his position in a government laboratory. His research interests are varied, including studies of primary cosmic radiation at high altitude by means of rockets, balloons, and satellites, and of x-ray astronomy. He has recently become Head of the High Energy Astrophysics Branch.

PETER MEYER ("The Electron Component of Cosmic Rays"), of the Enrico Fermi Institute for Nuclear Studies and the Department of Physics, University of Chicago, Chicago. Born in Berlin, he received the Ph.D. in physics from the University of Göttingen. He has held appointments at Cambridge and the Max Planck Institute. He is a member of four scientific societies and is a widely recognized authority on cosmic radiation and astrophysics. A session on cosmic rays without involving Dr. Meyer would be inconceivable. He has been a pioneer in measuring electrons in cosmic rays, particularly at a time when many scientists argued that there could be no electrons in cosmic radiation.

SERGE KORFF ("Cosmic Ray Neutron Studies"), of the Department of Physics, New York University, New York City. Born in Finland, he received the B.A., M.A., and Ph.D. in physics from Princeton. Like many of our contributors, Dr. Korff has a list of curriculum vitae a yard long, including senior advisor to the United Nations Atomic Energy Commission, research appointments in Bolivia, Chile, Brazil, the Arctic, the Antarctic, and Mount Wilson. He has received four international citations of merit, including Chevalier Légion d'Honneur. But of all these honors the one that pleases him most is that he is presently President of the American Geographical Society, which indicates how scientific research can be fun. Professor Korff started out in the neutron field a long time ago. In 1940 he and Bethe and Placzek wrote the first definitive paper on cosmic ray neutrons in the atmosphere, not only at high altitudes but also subject to boundary conditions at the earth's surface. The equations in that paper gave results and approximations that are still valid today.

ROGER W. WALLACE ("Neutron Exposure in Supersonic Transport"), of the Department of Nuclear Engineering, University of California at Berkeley. Dr. Wallace is also a physicist at Lawrence Radiation Laboratory. He left Baltimore, where he was born, and received the B.S. from California Institute of Technology and the Ph.D. in physics from the University of California. He is a member of the American Physical Society, the Radiation Research Society, and the Health Physics Society, and he is a specialist in neutron scattering. As long as ten years ago he and a group with him did some of the early definitive studies of the atmospheric cosmic ray neutrons as a function of altitude in our atmosphere. In the past two or three years there has been much concern about the problems that might be associated with supersonic transport flight. Such planes will fly at 70,000-80,000 feet; with the prospect of such altitudes a serious question arises: what are the radiobiological effects one might experience in such a flight? So a task force was developed under the sponsorship of the Atomic Energy Commission, and Dr. Wallace was instrumental in determining some of the physics associated with this problem.

JACOB KASTNER ("Effects of Soil and Water on Cosmic Ray Neutrons"), of the Radiological Physics Division, Argonne National Laboratory, Argonne, Illinois. Born in Germany, he earned the B.S. degree from the University of Manitoba and the M.A. and Ph.D. in physics from the University of Toronto. He was a National Research Council Canadian Fellow, a captain in the Canadian Army, and a research physicist at General Electric Corporation. He is a member of five scientific societies and a specialist in spectroscopy, radiation, radiological physics, and radioactivity.

RAYMOND GOLD ("Cosmic Ray Neutrons and Carbon-14 Production"), Head of the Experimental Physics Section of the Reactor Physics Division, Argonne National Laboratory, Argonne, Illinois. Dr. Gold was a student of Dr. Korff at New York University where he earned the B.A. and M.S. degrees. His Ph.D. in physics was awarded by the Illinois Institute of Technology. He has had a broad experience at General Electric, Armour Institute, the Naval Radiological Defense Institute, and Avco Corporation. He is a specialist in experimental nuclear physics and applied mathematics. The work reported here was performed under the auspices of the U. S. Atomic Energy Commission.

ADOLPH N. WITT ("Interstellar Dust"), of the Department of Physics and Astronomy, University of Toledo, Toledo, Ohio. Dr. Witt was born in Germany, had part of his training at the University of Hamburg, and he recently completed his Ph.D. degree at the University of Chicago. He is an astronomer.

ARMAND H. DELSEMME ("Cometary Nuclei"), of the Department of Physics and Astronomy, University of Toledo, Toledo, Ohio. Dr. Delsemme is a native of Belgium and received his academic education at the University of Liège. He has had tours of duty in various parts of the world, including the Belgian Congo and Paris. Since 1966 he has been Professor of astrophysics at the University of Toledo. He has published extensively in the areas of astrophysics and geophysics.

## ABBREVIATIONS

| | | | |
|---|---|---|---|
| A | ampere | in. | inch |
| Å | angstrom | KeV | kiloelectron volts |
| A.U. | astronomical unit | kg | kilogram |
| BeV | billion electron volts | km | kilometer |
| cal | calorie | kV | kilovolt |
| cm | centimeter | L | lambert |
| cgs | centimeter-gram-second | lb | pound |
| C | coulomb | m | meter |
| cu | cubic | $\mu$l | microliter |
| cc | cubic centimeter | mg | milligram |
| ° C. | degrees Celsius (centigrade) | ml | milliliter |
| ° F. | degrees Fahrenheit | mm | millimeter |
| ° K. | degrees Kelvin | MeV | million electron volts |
| eV | electron volt | MV | million volts |
| ft | foot | oz | ounce |
| G | Gauss | psi | pounds per square inch |
| GeV | giga electron volts | ster | steradian |
| gm | gram | V | volt |

# Index